CORAL REEFS

Look for these and other books in the
Lucent Endangered Animals and Habitats Series:

The Amazon Rain Forest
The Bald Eagle
Birds of Prey
The Bear
Coral Reefs
The Elephant
The Giant Panda
The Gorilla
The Manatee
The Oceans
The Orangutan
The Rhinoceros
Seals and Sea Lions
The Shark
The Tiger
The Whale
The Wolf

Other related titles in the Lucent Overview Series:

Acid Rain
Endangered Species
Energy Alternatives
Garbage
The Greenhouse Effect
Hazardous Waste
Ocean Pollution
Oil Spills
Ozone
Pesticides
Population
Rainforests
Recycling
Saving the American Wilderness
Vanishing Wetlands
Zoos

CORAL REEFS

BY LESLEY A. DuTEMPLE

Endangered
Animals &
Habitats

LUCENT BOOKS, INC.
SAN DIEGO, CALIFORNIA

Library of Congress Cataloging-in-Publication Data

DuTemple, Lesley A.
    Coral Reefs/ by Lesley A. DuTemple.
       p.   cm. — (Endangered animals & habitats) (Lucent overview series)
    Includes bibliographical references (p.) and index.
    Summary: Describes coral reefs and the problems they face, discussing how they are affected by human activities and their possible future.
    ISBN 1-56006-597-4 (lib. bdg. : alk. paper)
    1. Coral reefs ecology—Juvenile literature. 2. Endangered ecosystems—Juvenile literature.   3. Coral reef conservation—Juvenile literature. 4. Nature—Effect of human being on—Juvenile literature. [1. Coral reefs and islands. 2. Coral reef ecology. 3. Ecology.] I. Title. II. Series. III. Series: Lucent overview series.
QH541.5.C7D88   2000
577.7'89—dc21
                                                        99-38212
                                                            CIP

Copyright © 2000 by Lucent Books, Inc.
P.O. Box 289011, San Diego, CA 92198-9011
Printed in the U.S.A.

Contents

Introduction

Every ecosystem on Earth has unique characteristics, and some ecosystems can fairly be called more unusual than others. But no ecosystem on Earth is as unusual as a coral reef, which is the only ecosystem constructed by, and composed entirely of, animals.

All coral reefs are built by coral polyps. Coral polyps are small animals, typically no bigger than a pencil eraser. Yet these tiny animals manage to build enormous structures. The Great Barrier Reef in Australia can be identified from outer space and is the largest structure on Earth ever built by living creatures. Not even humans, for all their ingenuity, have been able to outbuild coral.

Coral reefs have been on Earth for millions of years, offering homes to more than 25 percent of all marine life. Although the oceans cover approximately 70 percent of the earth's surface, marine life flourishes in relatively limited habitats, of which the coral reef is particularly hospitable. The biodiversity of a coral reef is so great that scientists estimate that for every known species living on a coral reef, up to one hundred more species are not yet identified.

Limitless diversity

The biodiversity among the known species that inhabit coral reefs is enormous. Scientists estimate that more than 40 percent of all marine organisms spend some portion of their lives on, or around, a coral reef. Coral reefs, and their companion mangrove swamps and sea grass beds, are the nurseries of the ocean. Without coral reefs, many species of fish would be unable to grow to maturity. And without coral reefs and mature fish, there would be no multibillion-dollar fish industry. Scientists are also discovering that many marine organisms, including coral polyps themselves, produce cancer-fighting chemicals, and others possess antibacterial, antifungal, and anti-inflammatory characteristics. A coral reef is a veritable medicine chest, with most of the medicines not yet discovered.

Coral reefs also support thriving recreational and tourism industries around the globe. In the United States alone, more than 3 million people know how to scuba dive and a coral reef is the preferred destination for most divers. Within the surfing community, the best waves are usually found breaking over healthy coral reefs. And many of the world's most beautiful beaches and resorts are found on tropical islands fringed by coral reefs.

The wealth and diversity to be found in a coral reef is nearly limitless. Yet this magnificent ecosystem is now in trouble. What little is known about reef systems suggests that they are vital to the survival of the oceans, and what isn't

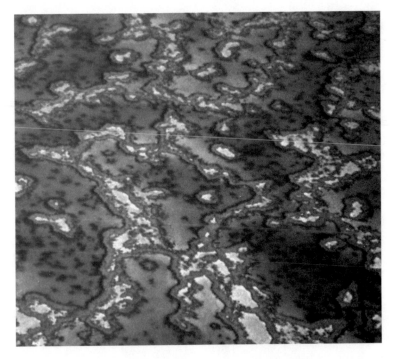

An aerial view of the Great Barrier Reef. Coral reefs house a diversity of life.

known about them staggers the imagination. The effect of the loss of even one coral reef ripples throughout medical, academic, and economic circles.

A lost resource

In the 1980s scientists began observing the bleaching (the expelling of the coral's algae, without which polyps cannot survive in the long term), and subsequent death, of coral reefs. At first, bleaching incidents were infrequent, but researchers soon reported widespread and frequent occurrences. To date, scientists estimate that at least 10 percent of the world's coral reefs are already lost; in some areas, more than 50 percent of some reefs are dead.

What is causing the corals to bleach? Any type of stress can cause a coral to expel its algae, or bleach. But prior to the 1980s bleaching incidents were very rare, and the af-

Divers flock to coral reefs to experience their natural beauty.

In recent years, changing conditions in the reefs have led to the widespread destruction of these rocky ecosystems.

fected coral usually obtained new algae and recovered. Since the 1990s, corals have been bleaching and not recovering.

Scientists know that several factors are causing coral reefs to bleach. Pollutants, sediment, excess nutrients in the water, and other factors are all stresses to coral reefs. Warming ocean temperatures are also a prime culprit. Coral reefs might withstand any one of these stresses, but in recent decades coral reefs have been subjected to numerous stresses at once.

Many reefs are also being physically destroyed by destructive fishing practices and overfishing. In many parts of the world, dynamiting a coral reef is the standard way to catch fish.

As we enter the new millennium, it is clear that human attitudes about the durability and value of coral reefs must change if the reefs are to survive. Governmental and private agencies alike are working to educate the general public and ensure that changes are made in how coral reef ecosystems are managed and maintained. With hard work, luck, and time, coral reefs may survive and thrive for another million years.

1

A Living Animal

CORAL. WHEN PEOPLE hear the word, they might think of a beautiful underwater reef teeming with life, a place where scuba divers fin delicately through schools of fish. Maybe they picture rosy-colored jewelry, hard and reflective. Or perhaps they think of a pure white "rock" in their aquarium. Whatever image the word *coral* brings to mind, it's rarely an animal. Yet that's exactly what coral is: an animal, and a very small one at that.

In the scientific classification of animals, coral belongs to the phylum of coelenterates—invertebrate, symmetrically round creatures including jellyfish and sea anemones. Coelenterates typically possess a saclike body with a mouth surrounded by tentacles. A single adult coral animal is called a polyp.

There are more than 2,500 species of coral in the world. All species of coral fall into one of two categories: reef-building corals (also called hard, or stony, corals) or soft corals. There are about 735 species of reef-building corals. The great majority, approximately 700 of these species, live in the Pacific and Indian Oceans, the remainder in the Atlantic Ocean. When people talk about a coral reef, they're really talking about hard coral. Hard corals secrete a stony shell of calcium carbonate around their body. The polyp itself is not hard, but the "home" it lives in is. These homes—lots of them—are what make a coral reef. Soft corals don't secrete a hard shell and, consequently, they don't form reefs. Soft coral is just that—soft—and can look a bit like seaweed.

Where coral is found

Although there are thousands of species of coral, nearly every species is found within a distinct global range. All coral, hard or soft, requires warm, clear ocean water. The water can't be colder than seventy-five degrees Fahrenheit during the coldest month, or warmer than eighty-eight degrees Fahrenheit during the warmest month. Hard coral requires fairly shallow ocean water and lots of sunlight. Consequently, it is usually found in water that is six to one hundred feet deep. Soft coral is found at approximately the same depths, but proliferates at the deeper end of the range—it can be found at six feet, but is more commonly found in deeper water. Many soft corals can tolerate slightly cooler water than hard corals, and a few can be found in temperate waters.

All corals can be killed by water that is too warm, too cold, too low in salinity, or too muddy, or by too much exposure to drying air or sunlight. In short, coral has very specific habitat requirements, and even small deviations can be lethal. The only area of the globe that meets all

A diver glides above a bed of soft coral.

Soft Corals

The majority of coral species found in the world are not reefbuilders. Reef-building corals are called hermatypic. Other corals, such as soft, black, thorny, and horny corals, are called ahermatypic corals.

Soft corals, though, are found all over coral reefs. Soft corals come in all shapes and sizes. Even their common names reflect their diversity: sea fan, sea plume, organ-pipe coral, and sea pansy. Whether red, orange, purple, or pink, soft corals are usually the most colorful fixtures on a coral reef. To a diver, soft corals can make a coral reef look like an underwater flower garden.

One reason why soft corals don't build reefs is because they don't have zooxanthellae. For unknown reasons, soft corals don't have the same partnership with these tiny algae that hard corals do. Zooxanthellae enable the hard corals to excrete the vast amounts of calcium carbonate necessary to form a reef. Soft corals secrete some calcium carbonate but not enough to form anything other than their own individual, flexible shapes.

Soft corals exist in an array of sizes, colors, and shapes.

these requirements is the region around the equator. All hard corals are found in the tropics, in a narrow band from the Tropic of Cancer to the Tropic of Capricorn, as are most soft corals.

Reef-building corals

Hard corals are not the largest category of coral species, but they are the best-known group. When people hear the word *coral,* they almost always think of hard coral. They think of a coral reef.

A coral reef is really a hard coral colony—millions and millions of tiny hard coral animals living together, forming one large coral reef. By reproducing itself, this tiny animal can transform a barren seascape into a unique ecosystem teeming with life and rich in diversity.

Double reproduction

Reproduction is the key to the formation of a coral reef. Building a coral reef many miles long requires a lot of polyps. To accomplish the task of reef building, nature has equipped coral polyps with not one but two methods of reproduction: sexual reproduction and asexual reproduction.

For sexual reproduction to occur, a male and female must breed, or mate. After mating, a fertilized egg is produced. According to marine biologist Thomas M. Niesen,

> The union of sperm and egg (fertilization) restores the full complement of genetic material, creating a new individual with a combination of the genetic characteristics of the parents. Such genetic variation may result in organisms better adapted to their environments or that are able to adapt to new environmental conditions.[1]

Hard coral is responsible for the building of a reef.

Sexual reproduction assures genetic diversity within a species.

In asexual reproduction, an animal reproduces itself—that is, an exact duplicate of itself—without mating. As Niesen points out,

> Non-sexual (asexual) means of reproduction are used by many species to take advantage of favorable—and rapidly changing—environmental conditions. Great increases in numbers can be achieved more quickly by asexual than by sexual means. . . . The coelenterates are masters of asexual reproduction.[2]

Asexual reproduction doesn't produce diversity, but it can produce large populations faster than sexual reproduction.

Spawning coral

Coral reproduces sexually through spawning. When coral spawns, large quantities of eggs and sperm are deposited directly into the surrounding water. Scientists unanimously agree that coral spawning is one of nature's most spectacular events. It usually occurs only once a year and on only one night.

A group of coral spawn in the waters off of western Australia.

Until the 1980s, scientists knew very little about coral sexual reproduction. Then a group of graduate students from James Cook University in Australia began studying the Great Barrier Reef, a twelve-hundred-mile-long coral reef off the coast of Australia. After months of underwater study, they were able to witness, and explain, how coral sexually reproduce.

Most coral polyps contain both sperm (male) and egg (female) reproductive cells. Both types of cells are formed deep in the polyp's body. Every year, as the earth's angle tilts closer to the sun, the ocean water warms up a bit from the increased sunlight. This increase in temperature stimulates the maturation of the coral polyps' egg and sperm cells. By the time spring arrives (the actual month varies depending on the location—in Australia, spring is in October, but in the Caribbean, spring is in April), the egg and sperm cells have ripened into a ball of reproductive cells called an egg bundle. At this point, the coral polyps are ready to spawn, or release their egg bundles into the water.

Researchers discovered that the coral of the Great Barrier Reef engaged in mass spawning, or spawning all at the same time. Scientists have since discovered that most coral species engage in mass spawning. There are exceptions; in the northern Red Sea, for instance, none of the major species of coral spawn at the same time.

Mass spawning makes good biological sense. According to author and biologist Rebecca L. Johnson,

> Researchers think that, by spawning all together, the corals overwhelm any nighttime egg-eaters with sheer numbers of egg bundles. When the water is thick with billions of reproductive cells, only a fraction will be eaten. The majority will survive simply because there are so many in the water at the same time. For these reasons, the mass spawning of so many corals seems to make sense.[3]

But how do the coral polyps know when to spawn, so that all of them spawn at the same time?

Most corals respond to the same signal when it comes to spawning: the full moon. During a spring full moon—whatever month that may be—the corals will spawn. It

On the night of a spawning, thousands of egg and sperm cells rise to the ocean's surface and break apart.

might be the first night of the full moon, it might be the third night, but it will be during a spring full moon.

As the full moon rises in the sky, the coral polyps begin releasing egg bundles. Each egg bundle is perfectly round, brightly colored—red, pink, or orange—and about half the size of a small gumball. The corals eject the egg bundles through their mouths, and the egg bundles rise to the surface of the water, where they form a thick, slightly oily, sweet-smelling layer. To a nighttime diver lucky enough to observe coral spawning, it looks as if an underwater blizzard is occurring. The spawning may last only a few minutes or continue for an hour. As suddenly as it started, it will stop—and the coral spawning is finished for another year.

When the bundles reach the surface, they break apart, releasing the egg and sperm cells. The sperm cells swim off in search of egg cells to fertilize. They never fertilize egg cells from their own bundle. Instead, they are attracted to egg cells from other polyps, either of their own species or from other coral colonies. With billions of cells in the water, it seems miraculous that sperm cells manage to find the right egg cells to fertilize. Scientists speculate that perhaps the egg cells release a special chemical to attract the right sperm cells, but this theory has yet to be confirmed.

Small beginnings

Once a coral egg is fertilized by sperm, it becomes a planula, or coral larva. Planulae are tiny, usually about half the size of a human infant's little fingernail. A planula is shaped a bit like a pear and has tiny hairs lining the sides of its body. Planulae swim by wiggling these tiny hairs. A planula might swim for hours, days, or even weeks after it is created, but eventually it will find a hard surface on which to attach itself. Once the planula attaches itself, it never moves from that spot for the rest of its life.

Once attached, the planula begins maturing into a polyp, or adult coral animal. As a polyp, the coral will develop a tiny slit of a mouth. Around this mouth, a ring of tentacles will begin to grow. By the time the planula has matured into an adult polyp—a process that may take several months—it is ready to start building a reef. It is ready to bud.

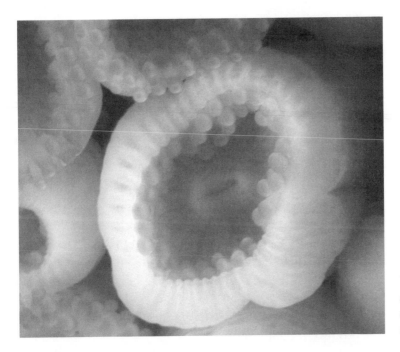

When planulae become adult-size polyps, they can begin to contribute to the reefs.

Budding

Budding is how corals reproduce asexually. If a planula settles on an appropriate spot and manages to mature into a polyp, it can be considered successful. As marine biologist Niesen points out,

> If a planktonic larval form (a planula) finds a suitable site for attachment and growth, chances are excellent that others of its species would also be successful in that area. If the organism can reproduce by asexual means, it can proliferate and maximize its use of the habitat in a relatively short period of time.[4]

A single planula essentially acts like a scout. By its very survival, it signals that the habitat is a good place to live. When a polyp is successful, it doesn't simply increase in size—it replicates itself and buds off another polyp. After its genetic material doubles, a polyp splits itself in half, and each half becomes a new complete individual. As Niesen explains, "This budding process continues and can eventually produce large coral heads consisting of thousands of asexually produced individuals."[5] The multiplying polyps are connected by coenosarc, a thin connective tissue that grows over the hard limestone between the polyps, like a membrane.

Budding can produce a lot of individuals in a single habitat. There will be no genetic diversity, however; each newly budded individual will have the exact genetic code of its parent. But there will be a *lot* of individuals without genetic diversity. In the case of coral, there will be enough individuals to begin the formation of a coral reef.

Hard shells and reef building

Coral polyps themselves, no matter how many, cannot produce a coral reef, at least not as humans know it. The polyps themselves are small, soft animals with tentacles surrounding their mouths. But as a polyp matures, it develops a protective shell about its body. Using calcium and other minerals contained in ocean water, each polyp manufactures and secretes a hard limestone skeletal cup, called a corallite. It is these hard exoskeletons—the corallites—that produce a coral reef.

Although polyps don't really increase in size, corallites can. As marine biologist James L. Sumich points out, "Periodically, the coral polyp withdraws its soft parts from the bottom portion of the corallite and secretes a partition. The partition provides a new elevated bottom on the corallite. Individual coral colonies may grow continually for centuries or even longer."[6] As the reef grows, coenosarc, the thin connective tissue, links the polyps and covers the hard limestone of the corallites.

The shape of the corallite will vary, depending on the species of polyp. It is the shape of the corallite that gives the colony, or the reef, its shape. Some hard corals look like giant brains. Others look like wavy sheets of paper or a huge cabbage. Still others look like antlers. Each of the approximately 735 species of hard coral currently known to exist in the world is capable of producing a differently shaped corallite. Many species produce corallites that look very similar to each other, while others can be wildly diverse.

This hard coral looks like a giant brain. As a coral reef is formed, different coral species give form to differently patterned reefs.

Corals can also change their shape, as the environment might require. In well-lit waters, some species grow in rounded and branching shapes. But in dim, poorly lit waters, the same coral will grow in wide, flat shelf shapes, which maximize its exposure to sunlight.

How polyps feed

Most coral polyps eat plankton, minute animals and plants that float passively or swim slowly in ocean water. Plankton, which also includes single-celled microscopic organisms, is the basis of the marine food chain. Many marine creatures—coral, clams, even gray whales—survive by eating the plankton found in ocean waters.

Generally, corals feed only at night. To feed, a polyp extends its body out of its corallite shell and opens its tentacles outward. The tentacles surround the polyp's mouth,

Polyps extend for their nightly feeding.

making the animal look like a small flower. A sticky mucous covering on each tentacle allows a polyp to capture plankton from the surrounding waters. As plankton becomes entangled in the tentacles, the polyp pulls the tentacles back into its mouth to digest the catch.

Plankton is an important part of a coral polyp's diet, but any polyp that relies solely on plankton for its food source will starve to death. Perhaps because a coral reef concentrates billions of polyps in one place, there just isn't enough plankton to go around. Whatever the reason, a hard coral polyp needs a food source other than plankton to survive.

Polyps in a symbiotic relationship

To get more nourishment, coral polyps have evolved a symbiotic, or mutually beneficial, relationship with a single-celled algae called zooxanthella. Zooxanthella can be found in ocean water, like plankton. But unlike plankton, zooxanthella has found a way to coexist with a coral polyp and to produce food for it. The polyp gets a built-in food source, and the zooxanthella finds a home, thereby saving itself from being eaten by other marine creatures.

Without zooxanthellae, there would be no hard coral polyps in the world because this added source of food is essential. There would also be no coral reefs, for polyps also depend on zooxanthellae to aid in the secretion of its calcareous corallite skeleton. These zooxanthellae actually live in the surface tissues of the polyp; they're not eaten and digested by the coral. Zooxanthellae also typically provide the color seen in corals. Zooxanthellae do not inhabit soft coral tissues, and, consequently, soft corals are incapable of secreting the hard limestone corallites necessary to construct a reef.

Because zooxanthellae are algae, they provide the coral with nourishment through the by-products of photosynthesis. Photosynthesis is the process by which plants create energy and nutrients from sunlight. By living in the coral's tissues, zooxanthellae use the waste materials the coral creates, and, in turn, the coral uses the waste products the zooxanthellae create. As marine biologist Sumich explains,

"In living corals, then, the zooxanthellae themselves are not digested, but sometimes as much as 90% of the organic material they manufacture photosynthetically is transferred to the host coral tissue. This is sufficient to satisfy the daily energy needs of several species of coral."[7]

Although plankton and zooxanthellae contribute greatly to the survival of coral, most polyps take advantage of any food that comes close enough to become entangled in their tentacles. Larger coral species, such as *favia* or *mussa,* which have large polyp bodies and long tentacles, even eat small fish. As Sumich points out, "Most coral polyps may be capable of harvesting organic particles as small as suspended bacteria, bits of drifting fish slime, and organic substances dissolved in seawater."[8]

The life span of coral

Every species of coral has a different growth rate. Branching corals grow the fastest, with the staghorn *acropora* growing at the phenomenal rate of fifteen centimeters a year. But faster growing colonies will also die faster, often breaking apart from their own weight or through the action of waves and boring organisms. Other species have sacrificed fast growth for endurance and long-range stability. *Porites* colonies are massive and grow only about nine millimeters a year, yet these colonies reach eight meters in height, thus making them nearly one thousand years old.

The life span of an individual coral polyp is difficult to determine. According to biologist J. E. N. Veron,

> By producing colonies composed of hundreds or thousands of individuals, corals are liberated from all the limitations of the single polyp. They can grow to a very great size, achieve great age, produce enormous quantities of larvae, grow fast enough to outmaneuver competitors and can construct plankton-catching sieves on a grand scale.[9]

Because of coral's ability to reproduce asexually, scientists have found it nearly impossible to study individual polyps in natural or aquarium settings. They don't really know what the life span of an individual polyp is. They

The Coral Reef in Your Medicine Cabinet

The value of a coral reef extends beyond the diversity of the unusual and beautiful creatures that flit through its sparkling waters. Scientists have begun investigating the marine life thriving in coral reefs for its medicinal value to humans.

More than six thousand unique chemical compounds have been isolated in marine organisms. Hundreds of these chemical compounds have provided drug leads, and others are in various states of testing. Already, the chemical compounds from several marine organisms are being developed as cancer-fighting agents. Others are being developed for their antibacterial, anti-inflammatory, antifungal, and antiviral properties. Scientists feel that they have discovered only a fraction of the useful chemical compounds available from marine organisms. Unfortunately, due to the destruction and degradation of coral reefs around the world, many potentially beneficial compounds are being destroyed before they can be discovered.

The destruction of coral reefs indirectly harms the academic institutions that research and develop new pharmaceuticals. The University of California alone has received more than $1.2 million in royalties from the development of patented pseudopterosins, a product made from the Caribbean sea whip that accelerates the healing process of human skin. The University of California also receives additional revenues from a major cosmetics firm that is now using pseudopterosins in its skin care products.

Coral reefs also supply the structural components necessary to repair human bones. According to the National Oceanic and Atmospheric Administration (NOAA,) "Interpore International manufactures bone graft material by converting the calcium carbonate endoskeleton of coral into calcium phosphate, or coralline hydroxyapatite, which closely resembles the physical and chemical structure of human bone." Unlike grafts of human bone, coral "bone" doesn't carry the risk of implant rejection or the transmission of infectious diseases such as hepatitis and AIDS.

With only a small fraction of coral reef life truly discovered, who knows what else is there—if it isn't destroyed before it's discovered.

think that the faster growing species, such as staghorn and other branching corals, might live to be ten to fifteen years old. That is the age when branching colonies can break apart from their own weight. But if one determines the age of an individual polyp by gauging the age of its colony, the larger, slower growing colonies might actually have individual polyps that are one thousand years old.

Once a coral colony has died, the calcareous reef structure remains. The reef structure itself will break down under wave action faster than, say, a rock. But depending on its location, dead reefs will remain in place for hundreds of years.

More than an animal

As an animal, coral is a fairly simple organism—a saclike body with a mouth and tentacles. It has an interesting symbiotic relationship with zooxanthellae, but many other creatures in the animal kingdom form symbiotic relationships. What makes hard coral a unique animal is that no other creature can do what it does. Hard coral occupies a unique and irreplaceable ecological niche.

It is hard coral's reef-building abilities that give it a unique place on this planet. For in constructing reefs, coral creates a habitat that is found nowhere else in the ocean or on the planet.

2

An Underwater City

SHAFTS OF SUNLIGHT flicker through the shallow crystal water. Jewel-like fish dart gracefully among swaying fronds. A shrimp scuttles about, ever alert for its next meal. Moving out of the shallows, the water grows deeper and cooler. Sunlight gives way to shadow, and a shark cuts sinuously through the depths, never blinking, never stopping. This is the world of the coral reef.

A coral reef is unique not just because it is an ecosystem that is created by an animal. Wherever a coral reef is formed, abundant life follows. Coral reefs are the oases of the ocean. According to author Douglas H. Chadwick, "Living coral reefs cover 360,000 square miles, an area slightly smaller than British Columbia (Canada), yet they host one of every four ocean species known."[10] By this reckoning, approximately 25 percent of all life in the ocean is found within a tiny geographical range—the range of the coral reefs.

Types of reefs

What exactly is a coral reef? True, it's really just a mass of coral polyps and their limestone skeletons. But a coral reef is more than just the sum of its polyps and their corallites. Each coral reef is different. Based on the location where the first polyp settled and the age of the colony, each coral colony creates a unique reef structure.

Although reefs can grow for hundreds of years, only the top portion of a reef—the polyps and coenosarc tissue—is alive and growing. The inner and lower layers of a reef are

A fringing reef, the most common type of coral reef, borders the shoreline of St. Thomas.

the skeletal remains—the limestone corallite structures—of corals that have died. These skeletal remains are compacted and fused as the live polyps continue reef building on top of them. Each reef exhibits a unique structure as it grows, but scientists classify all reefs into one of three categories: fringing, barrier, or atoll.

Fringing reefs

Among coral reef forms, a fringing reef is the most common shape. According to marine biologist Thomas M. Niesen, a fringing reef "extends seaward from shore, and is found surrounding islands and bordering continents."[11]

Fringing reefs always border a shoreline. Usually, they extend directly from the shoreline. But not just any shoreline will do, even within the narrow geographical band of the globe where coral reefs are found.

A vigorously growing coral reef requires a firm sea bottom (on which the first polyp will anchor), plenty of sunlight, clean water, and moderately high water salinity. Consequently, you won't find a fringing reef near the mouth of a river or where there is a lot of runoff from the land. The fresh water and suspended sediments in the water would make it impossible for coral to thrive.

Fringing reefs are a relatively young stage of reef development. Nearly every coral reef starts out as a fringing reef. As the reef grows and ages, it may change shape. Many Hawaiian reefs are fringing reefs. The largest collection of fringing reefs is found in the Red Sea.

Barrier reefs

Barrier reefs are found farther from shore. They are usually separated from the shoreline by a lagoon, or a shallow body of water enclosed between the shoreline and the reef. Sometimes, as in the case of the Great Barrier Reef in Australia, the lagoon may be miles wide, placing the reef miles from the actual shoreline.

The Great Barrier Reef in Australia is the largest structure ever built by living organisms.

Barrier reefs are older than fringing reefs. Typically, barrier reefs are also larger than fringing reefs. The most famous coral reef on Earth is a barrier reef. The Great Barrier Reef stretches for more than twelve hundred miles along the northeastern coast of Australia. Actually a collection of at least twenty-six hundred individual coral reefs linked together, it is the largest single structure made by any living creature on earth—not even humans have managed to create something as large as these tiny Australian coral polyps have.

Atoll reefs

Atolls take the longest amount of time to form and are therefore the oldest type of coral reef. An atoll is a ring-shaped reef with a lagoon in the middle of the ring and deep water surrounding the outside of the ring.

Atolls are found throughout the South Pacific, often hundreds of miles from any other island or land-form. To sailors, they are an accident waiting to happen, an obstacle in the ocean. But no matter their shape or location, they are still a form of coral reef teeming with life.

Volcanoes, coral reefs, . . . and time

For many years scientists were uncertain about how the three different reef formations came to exist. Oddly enough, the oldest theory, proposed in the 1830s, has turned out to be the correct one.

In the 1830s British naturalist Charles Darwin was serving aboard the HMS *Beagle,* an English sailing vessel commissioned to explore the world. The observations that Darwin made on this voyage, as well as the conclusions and theories that he subsequently developed, revolutionized the biological sciences and introduced the study of evolution.

During the voyage Darwin studied the reef forms of several oceanic islands. He hypothesized that most coral reefs were supported by volcanic mountains beneath the surface. Darwin also proposed that fringing reefs, barrier reefs, and atolls—rather than being entirely separate entities—were actually sequential stages of island reef development.

Darwin's theory was that newly formed volcanic islands and submerged volcanoes just under the surface were eventually populated by planktonic coral larvae. The coral settled and grew close to shore, just under the surface. As they colonized, they formed a fringing reef. Darwin also noted that the most rapid reef development occurred on the outer side of the reef because this was where food and oxygen-rich water was more abundant—so the reef was always extending farther from shore. As the reef grows, the weight of the expanding colony, as well as the enormous weight of the volcano itself, presses down on the sea floor and causes the island to start sinking. If the upward growth of the reef matches the sinking of the volcanic island, then the reef will be able to maintain its position in sunlit waters. If not, the reef will sink into darkness and perish.

Marine biologist James L. Sumich expands further:

> As the island sinks away from the growing reef, many of the massive corals left in the quiet waters behind the reef die. The corals are soon covered with reef debris and form a shallow lagoon . . . [in] what is now a barrier reef. With further sinking, the volcanic core of the island may disappear completely beneath the reef cap and leave behind . . . [an] atoll.[12]

Nearly 170 years later—with virtually no modifications—Charles Darwin's theory of reef formation still stands. Test drilling on several reefs has confirmed his theory of a volcanic base and the time progression involved in reef formation. As Sumich points out, "Two test holes drilled on Enewetak Atoll . . . penetrated over 1,000 m of shallow-water reef deposits before reaching the basalt rock of the volcano. For the past 30 million years, Enewetak apparently has been slowly subsiding as the reef grew around it."[13]

Although the common theory is that reefs and islands eventually submerge—even if it takes millions of years—reefs don't always sink because of their own weight. Sometimes they are submerged by a rising sea. Anytime the earth goes through a warming stage, glaciers and polar ice caps melt. As this water enters the ocean, the sea level rises. Throughout its history, the earth has experienced several fluctuations in sea level. The living coral reefs we

Charles Darwin hypothesized about the curious ability of coral reefs to form miles away from the nearest shoreline.

Plate Tectonics and Coral Reef Development

Charles Darwin correctly surmised the nature of fringing, barrier, and atoll reefs, but he was unaware that global plate tectonics also affect the development of a coral reef.

The earth is covered with a series of "plates" that fit together like a jigsaw puzzle. But because the earth's interior is molten and continually moves, the plates also shift a bit, which moves the islands and landmasses resting on them. The Hawaiian Islands chain and its surrounding coral reefs have been moved northwest from their original location by the movement of the Pacific plate. Scientists have noted that atolls at Darwin Point, at the northern end of the chain, appear to have drowned because they've been moved into deeper water faster than they can grow.

At Darwin Point, only about 20 percent of the calcium production necessary to build a coral reef is contributed by the corals. But even though the corals can't grow fast enough to keep the reef going, the reef still continues to some degree. The rest of the reef is formed from calcareous algae, calcareous marine organisms such as mollusks and sea urchins, and the skeletal remains of other reef inhabitants.

see today are the reefs that were able to keep up with the rising sea level. Others have perished beneath the rising waves through lack of sunlight.

Throughout the earth's history, reefs have come and gone. The city of Chicago stands on what was once a coral reef, and many of its buildings are made of gravel and cement formed from the crushed corals and shells that formed the reef 410 million years ago. But since the last ice age, about ten thousand years ago, most coral reefs have enjoyed a fairly stable environment. Throughout the centuries, many animals and plants have adapted to the habitat that is the coral reef.

Reef zones

A reef can be divided into zones. The part of the reef that faces into the lagoon is called the back reef. It's usually not much higher than the lagoon floor. The back reef then extends upward to the reef flat, which is a long flat area extending seaward to the reef crest. The reef crest is the highest point of the reef. On the other side of the reef crest is the reef front. The reef front faces the sea. Within the reef front are two subzones. The upper reef slope is directly below the reef crest. This is the zone where most reef life is found. Just below the upper reef slope is the lower reef slope. The lower reef slope is in deeper water and continues sloping downward until it reaches the seabed.

On Earth, coral reefs are second only to the rain forest in diversity of plant and animal life. At least two thousand different species have been observed in just one area of the Great Barrier Reef. Although most reef creatures are found in the upper reef slope zone, the whole reef is a hive of activity and life-forms. Each zone harbors different creatures and plants, and if the zones themselves didn't offer enough diversity, there are even three sets of population within the zones: a diurnal, or daytime, population; a nocturnal, or nighttime, population; and an in-between population that is active at dawn and dusk, when the diurnal and nocturnal populations are changing over.

We think of tropical seas as brimming with life and activity. But, compared with colder northern waters, tropical seas have very little life floating in them. Tropical waters are crystal clear *because* there is so little plankton, nutrients, minerals, and decaying plants suspended in the water—all of the things that marine plants and animals need to survive. A coral reef is like an underwater

Fish swarm in the Red Sea coral reef.

city, or an oasis in the desert. It's a bustling hive of activity because it provides perhaps the only source of food and shelter to marine creatures for miles around. Away from a coral reef, tropical waters are usually sparkling clear and empty.

A coral colony is a living thing and can support only a limited number of plants and animals that have the same shelter and nutrient requirements. Within the reef, space is limited and competition is fierce. A single pair of fish may live in only one particular place on the reef, above a coral boulder, for instance. They will defend their territory against other fish of the same species, keeping the spot only for themselves. Having different schedules—diurnal, nocturnal, and in-between—is one way that reef dwellers have adapted to make use of the limited resources.

Daytime on a coral reef

On a coral reef, the diurnal shift is the most active. That's because fish are the predominant life-form on a reef—

An abundance of life depends on the coral reef for food and shelter.

thousands of species of fish make their home on a coral reef—and most fish are diurnal, or active during the day. Of the approximately twenty-one thousand species of fish on earth, nearly 40 percent of them spend at least some time on coral reefs. Between one-half and two-thirds of all reef fish are diurnal. Angelfish, butterfly fish, triggerfish, clownfish, and barracuda are just a few of the many reef fish that are active during the day. With thousands of fish darting about, the activity level of a daytime reef resembles chaos.

Reef fish tend to be brightly colored. Although thousands of fish inhabit a healthy reef, within any given reef there are usually few individuals of any one species. Scientists believe that the dazzling colors may help the fish communicate and attract a mate. Some reef fish even change color when they are excited or threatened.

Other reef fish don't want to be noticed. Some, like the scorpion fish, are camouflaged. They sit quietly on the ocean floor, hidden among the seaweeds and coral chunks, waiting for unwary victims to unwittingly swim past their mouths.

Most diurnal fish possess good eyesight and color vision. Both of these qualities, plus good hearing and a keen sense of smell, help diurnal reef fish survive.

Many different types of plants live on coral reefs, although they're not the type of plant you'd find above water. Most plants that live on coral reefs are algae. Some algae, or too much algae, can be harmful to a reef. But many algae are a natural part of a reef ecosystem. Zooxanthellae are found in the coral itself, while other types of algae grow on the surface of the reef. Anchored to the reef and swaying with the current are different types of seaweed. But what most people call seaweed is actually a form of algae. Phytoplankton, a microscopic form of plant plankton, floats in the water surrounding the reef. All these plants form the basis of the reef's food chain. As the sun streams down through the crystalline water, through photosynthesis, the plants fill the water with enriching oxygen. They also provide homes and hiding places for reef inhabitants.

Coral Groomers

Within the reef, many creatures have evolved into specialized niches and diets. The triggerfish shoots a spray of water at a sea urchin's spines, causing it to overturn. Once its vulnerable underbelly is exposed, the triggerfish eats it. Long-nosed butterfly fish use their mouths, which are shaped like a pair of needle-nosed pliers, to grab worms, small shrimps, and coral polyps. The parrot fish, a large, colorful fish with a mouth remarkably similar to a parrot's beak, has one of the most unique niches on the reef: It eats the coral and recycles the reef.

A parrot fish's teeth are fused together, forming a strong mouth that works like a scraper. The fish frequently travel over the reef in schools, knocking off and chomping on chunks of coral. Parrot fish feed on algae that grow on dead coral as well as on zooxanthellae found in live coral. In this instance, living inside coral doesn't save zooxanthellae from being devoured by parrot fish.

The parrot fish scrapes off a mouthful of coral and swallows. A second set of teeth, located in the parrot fish's throat, grinds up the coral into fine particles that are excreted, creating coral sand for many of the world's tropical beaches.

One study in Bermuda found that for every acre of coral reef, one ton of coral skeletons were converted into sand each year— primarily through the feeding habits of parrot fish.

A parrot fish hides among the coral.

Dawn and dusk on a reef

As the afternoon sun sinks in the sky, the activity level of a reef changes. The fish sense the lengthening shadows and the frenetic activity that has been going on all day begins to slow. Fish become less interested in feeding and more interested in finding a place to spend the night. Many of these fish have a regular sleeping spot and start making their way to it.

Some diurnal reef fish can alter their color cells so they are no longer so brightly colored. Others simply look duller in the gathering twilight. But for most, the best defense is to simply disappear into the reef crevices. No matter what their color, they now form dark silhouettes against the setting sun—the perfect target for the predators that are moving onto the reef from the deeper waters below.

There are always predators—sharks, barracuda, groupers, snapper, and others—hanging about and inhabiting a coral reef. But during the day, many of them stay in the deeper waters of the lower reef slope. Others hang about the reef all day, not actively hunting, just waiting in the background for an easy opportunity or the consistent rush of activity that occurs at dusk and dawn.

Twilight predators usually have different vision than diurnal fish, even though they're out and active during daylight hours. They're not very good at distinguishing colors, but they're excellent at detecting shape outlines and movement. As the daytime fish make their way to resting places, the stream of movement presents the perfect target for many of these twilight predators.

Sharks can be found on the reef at any time of day. Although dawn and dusk offer perfect hunting opportunities, sharks have such a keen sense of smell that they can zero in on prey any time they choose. Vision is probably their weakest sense, but even sharks have catlike eyes that will mirror and magnify dim light, allowing them to hunt in the gathering gloom.

For some of these predators, dawn and dusk are the only times they're actively swimming about and hunting. It's their own particular time niche on the reef. Many of them,

As daylight fades on the reef, predators like these white tip reef sharks begin their hunt for food.

though, will continue to hunt after the sun goes down and the next reef population emerges.

Nighttime on the reef

About ten minutes after the sun sets, the reef is silent. Tired fish have settled down into crevices, and many of the larger predators have headed back down into the deeper waters beyond the reef. The calm is short-lived, though. Within fifteen more minutes, the reef will spring to life again.

The coral polyps themselves are nocturnal. All day long they have been tightly closed, holed up in their calcareous skeletons. Now, as darkness overtakes the reef, the polyps extend their bodies and unfurl their tentacles, sweeping the water for plankton and other nutritious debris.

Nighttime is also when coral polyps go to war. Fish and other marine creatures aren't the only ones with limited space on the reef. Coral polyps also compete for space, and sometimes neighboring coral colonies go to war. They may live peacefully side by side during the day, but at night, they attack. For instance, a fungus coral colony might exist next to a cactus coral colony. During the day nothing happens. But at night the fungus coral will send out poison darts to stab and digest its neighbor, the cactus coral, perceived as an enemy by the fungus coral, which will then expand into the space the cactus coral is occupying. During the day, white scars on the cactus coral will be the only evidence of the attack. Sometimes an attacking coral will completely overtake the weaker colony. Scientists who have witnessed these wars report that coral colonies only attack colonies of a different species. They never attack a different colony of their own species.

As the darkness deepens, other creatures emerge from the reef. Tube worms, sea stars, sea urchins, feather dusters, and others all come to life when the sun goes down. Invertebrates, or animals without backbones, form another large reef population. Brittle stars spread out their arms, positioning themselves to feast on plankton. Crabs and spiny lobsters crawl into crevices and patrol the ocean floor, ever alert for food. Shrimps clamber about on the coral itself, looking for edible debris and unwary prey. Nudibranchs, or sea slugs, appear to float in the dark water.

The moray eel is an example of a hunter whose skills are sharpest at night.

Moray eels emerge from the reef in search of their favorite prey, the octopus. Both of these creatures can be seen on the reef during daylight hours, but they are true creatures of the night. The octopus relies on its highly developed senses of taste and smell to find prey. The rims of the suction cups lining its eight arms are equipped with sensitive taste cells that can detect sweet, sour, and bitter better than the human tongue. An octopus can even tell a live clam from a dead one—without prying open the shell. Likewise, the moray eel relies on its sense of smell to hunt octopuses. Squeezing its snakelike body into cracks and tunnels, it can track an octopus into its den.

Coral Reefs: The Next Frontier?

For every marine organism living in a coral reef that has been discovered, described, named, and placed in an appropriate taxonomic category, scientists estimate that there are up to one hundred other species that haven't been discovered. The number varies based on the type of marine organism. Most large fish have been identified, but smaller groups of marine animals, such as flatworms or annelid worms, have not been cataloged or studied in detail. Undiscovered octopuses, in particular, are baffling scientists.

According to Mark Norman, an octopus expert at the University of Melbourne,

> We have new octopus species coming out our ears on coral reefs throughout the Indo-West Pacific. To give you an idea of the scale of the issue, we have recently recognized 40 species in northern Australian waters, of which more than 30 are new to science. Examination of animals from New Caledonia found 30 species, of which 26 are new. We have just published a paper on Philippine octopuses recognizing 26 species of which only 4 were previously correctly identified, and we have at least 30 new species occurring within Indonesian waters.

Many marine biologists are now turning their efforts toward cataloging coral reef life-forms, hoping to do so before unknown creatures disappear.

Soon, a whole new group of fish appear—the ones that feed on invertebrates. Squirrelfish, grunts, and drumfish all head out to feed in the lagoon or deep waters of the reef slope. These fish have been sleeping all day in the crevices and tunnels of the reef. At dawn and dusk they awake and essentially change places with the diurnal fish. Grunts stir up the sand, eating sea worms, shrimps, and anything else they can find. Squirrelfish cruise the steep wall of the reef slope looking for zooplankton, jellyfish, or larvae of other sea creatures.

Scientists think that long ago, when dinosaurs were on Earth, these nocturnal fish inhabited the reef and were diurnal. As the dinosaurs died out, new fish species emerged, many of them with better eyesight and feeding abilities. The new fish were better equipped to compete for space on the reef. Gradually, the more ancient species—the squirrelfish, grunts, and drumfish—were crowded out until they were confined to the nocturnal feeding shift for survival.

Other creatures also use the coral reef. During the day sea turtles swim gracefully in the warm reef waters. By night, they swim over the reef, crawling onto the sandy beach to lay their eggs. For all of these creatures—parrot fish, angelfish, shrimps, spiny lobsters, sea turtles, grunts, sea stars, and thousands of other species—the coral reef is home. Without coral reefs, all of this life would cease to exist. For millions of years, coral reefs have evolved and supported the greatest amount of marine life within the smallest amount of space. But now, coral reefs—and the life forms that inhabit them—are in trouble.

3

Environmental Threats to Coral Reefs

ALTHOUGH CORAL HAS been on Earth for millions of years, it is more fragile than its survival record would indicate. For a creature that has survived nearly every natural disaster, including the extinction of the dinosaurs, coral is not all that adaptable. And when it does adapt, it doesn't do it quickly. Today, changes to its environment are occurring faster than coral can adapt. Nearly all of these changes and problems are anthropogenic, or caused by humans.

Bleaching reefs

When people visit the tropics, one of the first things to strike them is the color of the water. A brilliant turquoise blue shifts effortlessly into opalescent green; tropical waters change colors faster than a chameleon. But frequently it's not the water itself that's so lovely—it's the vibrantly colored coral on the bottom, which is viewed through completely transparent tropical water. Healthy coral is what provides color and life to tropical waters.

When coral is unhealthy, or stressed, it will bleach itself. When a reef is bleached, it loses its color and becomes white and skeletal. A bleached reef looks dead and often does die. The phenomenon of coral bleaching was first observed about a century ago and was considered at the time to be very unusual behavior.

Coral bleaching is caused by coral polyps' expelling their zooxanthellae partners. The zooxanthellae provide both nourishment and color to their host polyps. But when coral polyps become sick or stressed, they expel the zoothanthellae. If the stressful conditions subside within a few days or weeks, the coral will accept new zooxanthellae and resume their symbiotic relationship. But if the stressful conditions continue for weeks, months, or even years, the coral will most likely die of starvation. Without zooxanthellae, the polyp is unable to obtain enough nourishment from plankton, small fish, and floating debris to survive on a long-term basis. Under stressful conditions, keeping the zooxanthellae requires more energy than the zooxanthellae provide.

If the stressful conditions are intermittent, and the coral accepts new zooxanthellae only to repeatedly expel them,

When bleaching takes place, coral loses its color and begins to die.

the polyp becomes so weakened that even a short bleaching will kill it. Through repeated short-term bleachings, or through a single long-term bleaching, an entire reef can kill itself.

Major coral bleachings started occurring around the world in the 1980s. The first episode was in 1980. Three years later, in 1983, another major bleaching occurred. A third occurred in 1987, with yet another in 1990. And the bleachings were occurring everywhere, not just in a single location. Even more troubling to scientists was the fact that each bleaching episode appeared to be more severe and last longer than the previous one.

Stressed coral

Many things can stress a coral polyp and cause it to bleach itself. According to the National Oceanic and Atmospheric Administration (NOAA), "Just like people, corals must endure stress from many different sources in order to survive. Some of these stresses are natural, such as black band disease or grazing by other marine organisms. However, the greatest threats to coral survival are stresses which are anthropogenic, or 'created by humans.'"[14]

Anything that negatively affects the quality of the water in which coral lives can cause coral to bleach itself. Pollutants, sediment and nutrients, leakage from oil drilling operations, agricultural runoff, diseases, and changes in the water temperature are all possible causes and contributing factors to coral bleaching. Unfortunately, in some parts of the world coral are subjected to many of these stresses simultaneously. According to a report by NOAA, "Today, 10% of the world's coral reefs have already been destroyed, and scientists predict that we will lose an additional 30% within the next twenty years as the intensity and frequency of these anthropogenic stresses increase."[15]

Coral and global warming

The earth's climate is subject to fluctuations in temperature. Currently, the earth appears to be in a phase of rising temperatures, which is bad news for corals around the

world. When water temperatures rise, corals become increasingly stressed. Many corals already live in waters that are very near their upper temperature limit, so a change of a degree or two can be deadly. The earth's temperature has warmed before, naturally, but the current warming phase appears to be almost entirely anthropogenically driven—the warming of the earth and its seas is being caused by humans.

Global warming may be causing large areas of coral reef to die.

The current warming of the earth is referred to as global warming. Experts do not agree on the causes and severity of global warming, but many feel that the burning of fossil fuels such as coal and oil are altering the protective layer of atmosphere that surrounds the planet. Sunlight penetrates the atmosphere, warming the earth and the oceans, but excess heat cannot escape because it is unable to penetrate the altered atmosphere. Over time, the earth is gradually warming and ocean temperatures rise.

Rising water temperatures stress coral and cause them to eject their zooxanthellae, thus bleaching themselves.

The Crown of Thorns

Coral reefs in the Pacific and Indian Oceans are subject to attacks by a coral-eating sea star, the crown of thorns (COT). These large sea stars grow to nearly a foot in diameter (thirty centimeters), about the size of a frisbee, and devour coral polyps. For unknown reasons, every few years the population of these prickly predators increases drastically. Scientists have speculated that perhaps the decline of tritons, a large snail that is one of the few COT predators, has caused an increase in the population. Tritons are prized by shell collectors; and, due to overharvesting, their numbers are dwindling.

When one COT arrives on a reef, others soon follow. When one of these large sea stars eats, it releases a chemical that attracts other COT. So they gather in great numbers, eating huge portions of a coral reef. In the Maldive Islands, scuba divers are asked to pry the COT off the reef in an effort to control the damage. At only one resort in the Maldives, divers removed more than eighteen thousand in one year. Since the 1990s, the Great Barrier Reef in Australia, as well as reefs in Fiji, Indonesia, Mauritius, Maldives, the Philippines, the Solomon Islands, Samoa, South Africa, Vanuatu, and the Red Sea region have come under attack.

When COT attack, they distend their stomach over the coral and release a chemical that dissolves the coral, turning it into a soupy pulp. The COT then slurps it up through its mouth. The sea stars travel over a reef, literally scouring it clean. It may take a reef twenty years or more to recover from a COT attack.

A crown of thorns sea star can seriously damage a coral bed.

According to Billy Causey, superintendent of the Florida Keys National Marine Sanctuary, "Temperatures of 86–87.8 degrees F are common triggers. If these elevated temperatures continue for one to two months, the results can be lethal."[16] During the summer of 1997, sensor buoys in the Florida Keys detected temperatures ranging from 86 degrees Fahrenheit to an extraordinary high of 91.4 degrees Fahrenheit. That summer Causey said, "This year's bleaching is beginning to look severe. In some spots we have seen as much as 70 percent of an area affected. Another indicator of the severity is that bleaching is occurring in the nearshore waters in species that generally are more adept at adjusting to temperature fluctuations."[17] As of June 1999, the coral reefs of the Florida Keys had still not recovered, and many species were still exhibiting bleaching.

Many scientists feel that global warming is currently the greatest threat to the world's coral reefs. But global warming is only one threat that coral reefs face.

Coral and pollution

Corals are particularly susceptible to chemical pollution. According to a NOAA report, "Corals absorb chemicals across their outer tissue layer, so whatever contaminants are suspended (floating) in the surrounding water will soon be absorbed into the coral itself."[18]

Any amount of pollution in waters inhabited by coral causes harm to coral. In low concentrations, pollutants may inhibit reproduction and growth in coral. In high concentrations, pollutants contribute to coral bleaching and the subsequent death of the reef.

In recent years scientists have been finding concentrations of heavy metals in coral colonies. Heavy-metal pollutants include lead, mercury, zinc, copper, tin, cadmium, iron, aluminum, and other elements. All heavy metals are toxic, or poisonous, to living creatures. Some of these heavy metals occur naturally, but the majority of heavy metals found in nearshore waters are entirely anthropogenic.

Some of these heavy metals come from offshore oil drilling projects, entering the water when drilling fluid spills into the water. They can also be found in anticorrosive paints that are used to protect ships and coastal structures. But scientists feel that most of these heavy-metal pollutants are coming from land-based sources such as mining and smelting operations, oil refineries, power plants, and other industries. The pollutants are entering storm drains and sewage systems, which then flow into rivers and other natural drainage systems, finally entering the sea, where they stress, or kill, coral colonies.

According to NOAA,

> The biggest problems with heavy metals are that they are persistent (remain in the environment unchanged for years), and they bioaccumulate (increase in concentration as they go up the food chain). So, in addition to posing a threat to the health of coral reefs around the world, toxic metals also pose a long-term public health risk, especially for those human populations that rely on fish for protein.[19]

Sediments and nutrients

Anything that diminishes water clarity or exposure to sunlight can stress and cause coral to bleach or slow its reproduction and regeneration. In many parts of the world, but specifically in the Caribbean, sediment and added nutrients in the sea have produced this response. Sediment usually consists of tiny particles of earth, or dirt. The particles remain suspended in the water, causing it to look cloudy, then sink to the bottom from their own weight. When they finally sink and settle, they can cause the bottom to become muddy. Nutrients can be anything that can be consumed and cause something to grow. For algae, many things can be a nutrient: fertilizers, sewage, even chemical pollution.

In the last ten years algae have overgrown many coral reefs, transforming these exquisite ecosystems into barren

Sunlight deprivation due to clouded water can turn coral reefs into barren algae-covered rocks.

algae-covered rocks. Extra nutrients in the water are causing normally present algae to grow at astounding rates. Algae thrive on the nutrients and, consequently, grow faster than the coral. As algae fill the water and cover the reef, they form a perfect trap for sediment. The sediment shades, then sinks, smothers, and kills the coral. Where are all of these nutrients and sediments coming from? Coral researcher Philip Dustan of the University of Charleston, in South Carolina, thinks he has part of the answer. Dustan points out that

> ecosystems on land are naturally "conservative." They don't easily give up elements to the sea. Whatever sediment and nutrients are dislodged tend to get trapped and contained. Coral reefs have adapted to the resulting clear water and low nutrient levels. But the modern processes of development have severely disrupted the land's capacity to hold onto elements. Agriculture, urban development, forestry and other practices have made it easy for rains to wash sediments and nutrients out to sea.[20]

The fact that reefs close to population centers suffer more algae and sediment damage than isolated reefs gives credence to Dustan's theory. But locating the actual source of all of these nutrients and sediments is difficult. For instance, in the Florida Keys pollution from numerous towns and cities seeps into the sea through a complex network of canals, streams, rivers, and coastal bays. The nutrients leach into the ground from septic tanks, sewage, lawns, and vacant lots, slowly entering the watershed network all the time, but even more quickly when it rains.

Sometimes the sediment and algae damage comes from sources that are nowhere near the coral reef. Studies conducted by Gene Shinn, a federal biologist working at the Center for Coastal Geology in St. Petersburg, Florida, indicate that Caribbean coral reefs are even being damaged by iron-rich dust from drought-stricken Africa. In the 1970s, as desertification began to spread across northern Africa, measuring stations in the Caribbean also began recording large increases in dust as well as disease outbreaks. In 1987, the year corals all over the Caribbean bleached, the dust levels also peaked. According to an article in the Winter 1999

Mangrove Swamps and Sea Grass Beds: A Healthy Partnership

The continuing health of coral reefs is also contingent on the continuing health of two other habitats: mangrove swamps and sea grass beds.

Mangrove swamps are semiaquatic wetlands where mangrove trees grow partially in and partially out of the water. Mangrove swamps are found along tropical shorelines. They're important to coral reefs because they act as strainers, filtering out sediment, pollutants, and other land runoff before they enter the water. The root system of a mangrove swamp also acts as a nursery for young fish and other juvenile marine creatures before they grow up and move out onto the reef.

Sea grass beds are underwater lawns where many reef-dwelling creatures feed. Sea grass beds are usually found in the lagoon portion of the reef system. Like the mangrove swamps, the sea grass beds also help to filter out sediments and pollutants as well as stabilize the sand and lagoon floor.

Scientists consider all three habitats to be integrally tied together: mangrove swamps, sea grass beds, and the coral reef. All three must be healthy for there to be a thriving coral reef ecosystem.

Mangrove swamps help support life in the coral reef by filtering out impurities in the water before they reach the reef.

newsletter for the Coral Reef Alliance (CORAL), "Iron in the dust stimulates the growth of algae which damage the reefs, and the dust also contains bacterial spores."[21] Marine biologist Richard Barber of Duke University thinks the situation needs global attention: "If the dust proves harmful to corals, efforts to reverse desertification in northern Africa should increase."[22]

Coral reefs and coastal development

As human populations burgeon around the globe, housing and recreational development of coastal areas also burgeon. Besides increasing sewage, fertilizer, and other runoff problems, coastal development can affect coral in more direct ways.

When a section of coastline is developed, the altered shoreline can affect wave action and cause the reef to experience changes—none of them beneficial. In other cases,

As coastal areas become more developed, problems with sewage and other human waste products will continue to plague the coral reefs.

reefs are actually destroyed to make way for piers and marinas. When coastal developments are constructed, harbors and channels are dredged, causing heavy metals and other pollutants that have been trapped in the bottom sediment to float freely, which further damages the coral reef.

Coral reefs and natural disasters

Humans aren't the only causes of coral reef devastation. Throughout the earth's history, natural occurrences have also wreaked havoc on coral reefs.

Hurricanes are, by far, the greatest natural destroyers of coral reefs. These gigantic storms, indigenous to the tropics, create winds up to two hundred miles per hour that whip up enormous waves. As the waves batter a coral reef, entire chunks of the reef are ripped off and swept away.

Hurricanes and other natural disasters can damage coral reefs.

Heavy, prolonged rains can also damage a reef by putting so much fresh water into the ocean that the salinity content lowers. Heavy rains can also cause rivers to swell, consequently dumping abnormal amounts of muddy water, silt, and sediment into the ocean—which subsequently clog and smother the reef.

And even though most coral reefs are located in tropical regions of the globe, these areas can still experience unusually cold or hot periods. If a prolonged cold snap occurs, the reef will be damaged just as surely as it is damaged by prolonged warmer temperatures.

Coral reefs and the coastal zone

Although coral reefs can certainly be damaged or destroyed by natural events, their greatest threats still come from humans. One reason why corals suffer so much from anthropogenic problems is because many reefs are found close to human settlements. Coral reefs require relatively shallow water, so most of them are located in what is referred to as the coastal zone. Unfortunately, around the world the coastal zone is the most heavily populated area.

According to a general report issued by the United Nations,

> Over half of the world's population lives within 60 kilometers (37.5 miles) of the coast and this could rise to three-quarters of the world's population by the year 2020. Because the terrestrial portion of the coastal zone hosts such large human populations, it is also the focal point of various land-based human activities such as urban development, agriculture, manufacturing, energy production, tourism, aquaculture, mining, forestry, and transportation.[23]

Reefs within an inhabited coastal zone are not only subject to the indirect effects of human activities, they're also subject to direct human interference. For many humans around the world, coral reefs are a workplace as well as a source of recreation. These direct human activities are causing further harm to coral reefs.

4

Fishing and Recreational Threats to Coral Reefs

WITH SO MUCH of the world's population living near coral reefs, direct damage from human activities is a growing concern among the world's scientific community. According to a recent State of the Reefs report put out by the Coral Reef Initiative, a consortium of nations around the globe,

> Coral reef ecosystems are under increasing pressure, and the threats are primarily from human interaction. Coral reef ecosystems at greatest risk are in South and Southeast Asia, East Africa, and the Caribbean. However, people have damaged or destroyed reefs in over 93 countries. . . . Technology also allows humans to exploit the reef with mechanical dredges, hydraulic suction, dynamiting, and large-scale poisoning.[24]

Blast fishing

Throughout Asia and the South Pacific, where many of the world's coral reefs are located, one of the greatest dangers to coral reefs is fishing. For centuries within this region, local fishermen have relied on reefs for their livelihood. But with overfishing, the worldwide deterioration of fish stocks, and the declining health of coral reefs, many of these fishermen have resorted to blast fishing in an effort to maintain their lifestyle.

Blast fishing involves dynamite or rudimentary explosives that are thrown from boats or detonated from a safe distance. When the blast goes off, huge numbers of fish and other marine creatures are killed, and the physical structure of the reef is reduced to rubble—effectively destroying the entire ecosystem. When the dead marine creatures float to the surface, the fishermen scoop them up in nets. It can take hundreds of years for a reef to recover from the decimation of blast fishing.

In Indonesia, the Komodo National Park encompasses three large islands and many other smaller ones. Although the park is known for the legendary Komodo dragons, it is also home to one of the world's most diverse marine ecosystems. Within the 173,000 hectares that comprise the park, more than 1,000 species of fish, 253 species of reef-building coral, and 70 sponge species can be found. Additionally, dolphins and whales frequent the waters around the park.

To date, local fishermen have destroyed more than 52 percent of the park's coral reef system, solely through blast fishing. Unfortunately, this phenomenon is not confined to

In response to increasing economic pressure, many fishermen in Asia and the South Pacific have resorted to blast fishing, which threatens to destroy the entire coastal reef ecosystem.

Saving Brazil's Unique Coral Reefs

Brazil's coral reefs run for three thousand kilometers along the Brazilian coast, from southern Bahia to Maranhão state in the north. These reefs are unique because they are the only coral reefs found in the entire South Atlantic region. The reefs are isolated from their Caribbean counterparts by the enormous amounts of silt-laden freshwater that pours from the Amazon and Orinoco Rivers. The freshwater effectively forms a twenty-seven-hundred-kilometer barrier, preventing the two different coral populations (Caribbean and Brazilian) from intermingling.

The isolated location of the Brazilian coral reefs has contributed to the evolution of unique marine organisms found nowhere else in the world. Of the eighteen species of hard coral that are found in Brazil, ten of them are endemic, or found nowhere else. This gives Brazil the highest proportion of endemic coral species in the world. Some of the species, such as *Favia leptophylla, Mussusmilia hartti,* and *M. braziliensis,* show archaic characteristics that, to date, have only been found in fossil form. Should any of these species become extinct, it is highly unlikely that they will ever be found on another coral reef.

Scientists' concern for the unique Brazilian reefs is intensified because there is currently no legislation aimed at protecting them. Coastal development, tourism, and destructive fishing practices are but a few of the stresses Brazilian corals face. The International Coral Reef Initiative is working with the Brazilian government to try and implement protective measures for sustainable practices around the reefs.

Komodo National Park. Blast fishing is a widespread practice throughout Indonesian and Asian waters and is greatly contributing to the destruction of coral reefs in the area.

Fishing with poison

Blast fishing is not the only fishing threat Southeast Asian and Southern Pacific reefs face. Poisoning is another large problem. According to CORAL,

> Cyanide, one of the most toxic poisons known, is being used to catch live fish in the South Pacific and Southeast Asia. Fishermen stun fish by squirting cyanide into the reef areas where these fish seek refuge. *They then rip apart the reefs with crowbars to capture disoriented fish in the coral where they hide.* In addition, cyanide kills coral polyps and the symbiotic algae and other small organisms necessary for healthy reefs.[25]

Cyanide fishing used to be confined to the aquarium trade. Approximately 85 percent of all marine aquarium fish are captured in Philippine and Indonesian waters. The other 15 percent are captured in the Caribbean, Red Sea, or Indian Ocean. Nearly all of these fish are captured using cyanide, and more than 60 percent of these cyanide-captured fish are purchased by Americans for their aquariums. Fishermen enter the water armed with squirt bottles of sodium cyanide poison. They squirt the poison at individual fish, or schools of fish, then scoop up the stunned fish in nets. Divers also use squirt bottles of liquid dish soap (equally poisonous to coral reefs and marine creatures) to flush out spiny lobsters and other marine creatures from their hiding places in the reef. Although cyanide fishing does not have a direct effect on the physical structure of the reef, like blast fishing, only a small amount of poison—cyanide or dish soap—will kill living coral. Without living coral, the reef—and the entire ecosystem—die.

But in the last ten years another market—besides the aquarium trade—has developed, and it relies almost solely on cyanide for the capture of fish. Nowadays, the demand for live fish in luxury Asian restaurants, particularly in Hong Kong and Japan, is driving the practice of cyanide fishing. To meet the demands of Asian restaurant goers, each year an estimated 330,000 pounds of cyanide are being sprayed on coral reefs in the Philippines alone. As Philippine reefs are devastated, practitioners move to remote, pristine coral reefs such as those in eastern Indonesia, Papua New Guinea, and other countries in the western Pacific.

According to NOAA,

> Because this reef-destructive fishing method is relatively efficient, it increases the likelihood that target fish species, and other species, will be overfished. Thus cyanide fishing threatens—directly through overfishing and indirectly through habitat destruction—the viability of the coral reef fisheries upon which hundreds of thousands of small-scale fishermen around the world depend for animal protein and economic livelihood.[26]

Although cyanide fishing is illegal in most nations, very few governments monitor or enforce the restrictions against cyanide use.

Marine groundings and coral

Many coral reefs are located close to major shipping lanes. With large ships traveling so close to coral reefs, it's not surprising that ships can frequently run aground on the reefs. A ship grounding can cause the same type of damage to a reef that blast fishing does—the reef is reduced to rubble. Throughout the world, groundings occur more frequently than anyone would like. Besides the loss of a valuable ship, a valuable resource—the coral reef—is also damaged, sometimes irreparably. Even protected marine sanctuaries are subject to this type of disaster.

A barracuda swims among the wreckage of a vessel that collided with a coral reef.

In the Florida Keys National Marine Sanctuary, several ship groundings have occurred since the 1980s, even though the area is a protected marine sanctuary. In 1984 the four-hundred-foot freighter *M/V Wellwood* ran aground on a coral reef, essentially reducing a vibrant coral reef into an "underwater parking lot strewn with limestone rubble."[27] In 1989 another large ship ran aground in the same sanctuary, destroying more than sixteen hundred square meters of formerly pristine

Reconstructing Coral Reefs

 When ship groundings destroy or damage a coral reef, the only solution used to be to wait one hundred years for the reef to repair itself. But in the Florida Keys National Marine Sanctuary, scientists are working on ways to reconstruct broken reefs.

An extensive salvage operation is currently underway to repair the damage caused by the grounding of the containership *Contship Houston,* which ran aground in the sanctuary on February 2, 1997. There, workers glue the fractured coral heads back together and right the sponges and corals that were overturned during the accident. Restoration work is also underway at other grounding sites in the Florida Keys.

Scientists often start by filling in the blowholes created by the ship with precast concrete blocks and limestone boulders to try to create the structure of the reef. These sturdy structures give larval organisms, such as coral and sponges, a firm surface on which to attach. The scientists then actively transplant sponges, hard and soft corals, and other organisms into the area. By deliberately "seeding" the area, biologists hope to jump-start the redevelopment of the reef.

Scientists also propose applying these techniques to remove coral from harm's way when all else fails. If successful, rather than being destroyed to make way for a new pier, a coral reef could be transplanted to a hospitable location. Evidence indicates that damaged coral in Mexico, Thailand, and Hawaii has responded positively to these limited salvage operations.

coral reef. In response to these groundings, and to minimize damage to the world's second largest barrier reef, the U.S. Congress designated the Keys as an "Area to Be Avoided" by any vessel longer than fifty meters.

Unfortunately, this measure has been unable to eliminate groundings in the sanctuary. Again, in February 1997, yet another large freighter ran aground.

Even a healthy coral reef that suffers this type of damage has trouble rebuilding. It may take a reef decades, even one hundred years, to recover from the effects of a ship grounding, and some damage is irreparable.

Divers: The problem or the solution?

When researchers first began reporting signs that coral reefs were in serious trouble, about twenty years ago, one of their first theories was that scuba divers were causing much of the damage. Once considered a luxury sport for the rich and daring, scuba diving has become a mainstream activity. Also once considered a sport for men only, improved technology has attracted many women and teenagers to the sport in the last decade. More than 14 million Americans have tried scuba diving, and 3.5 million of them dive on a regular basis.

In the early days of scuba diving, careless divers damaged portions of coral reef, including this section of the Great Barrier Reef.

For most of these millions of divers, a coral reef is the preferred place to dive. The diversity and physical beauty of coral reefs offer a diving experience unmatched by other locations. Divers can journey to a coral reef for a diving vacation, go diving in the same spot every day, and still see something new and different every time they dive. The sheer number of people scuba diving on coral reefs suggested to scientists that divers were causing the damage to coral reefs.

Thirty years ago, early in the history of scuba diving, divers may well have contributed to the destruction of coral reefs. It was common practice for a dive boat to motor out to a coral reef, toss out its anchor, and let divers jump unsupervised into the crystalline waters. In the process,

coral heads were damaged by anchors scraping and gouging them and by careless divers kicking and breaking them. Frequently wildlife also suffered, either inadvertently or intentionally, because scuba diving also opened up a new way of fishing. Divers could enter the realm of marine life and capture specimens by hand and by spear fishing, in some cases threatening to deplete stocks of some species as game or trophies. By the mid-1980s, though, nearly every diving operation around the world recognized that divers needed to be educated about the environmental impact of recreational diving or there would be no great places left to dive.

A marine patrol scuba diver at Lode Key National Park advises other divers to avoid touching the coral reefs.

Divers can indeed harm coral reefs, but the amount of damage done by divers, even with millions of people diving every year, is negligible compared to the global impact

of overfishing; destructive, poisonous fishing; and the dumping of sewage, fertilizer, chemicals, and sediment into the ocean. Divers can physically help protect reefs, but their real power to help preserve reefs is economic. Each year the diving industry generates more than $1.7 billion in tourism revenue. In many foreign areas, diving tourism is the primary source of revenue and employment. Divers have a direct interest in keeping coral reefs alive and healthy, but they also have the economic power to implement changes in the way coral reefs are managed and regulated by governments dependent on tourism dollars.

By the 1990s divers had realized that coral reefs needed help and also that, as a group, they had power. Divers began forming organizations to educate other divers and the general public as well as to raise funds to save many coral reefs. One of the largest dive organizations was formed in 1993. According to NOAA,

> In 1993, a group of scuba divers that wanted to help keep coral reefs alive formed a non-profit organization called "The Coral Reef Alliance" (CORAL). CORAL's education and conservation programs around the world have expanded beyond the dive community . . . to bring its message to school children, aquarium visitors, movie viewers, and many others who may not have much information about coral reefs.[28]

Recognizing that coral reefs are in serious danger all over the world, diving organizations have taken an activist stance. By raising public awareness of the problem, they are working hard to help save a valuable global resource. As CORAL frequently states in its literature, "Divers who care for the reefs are not the problem. Responsible divers are part of the solution."[29] Along with divers, many other global organizations are coming together to try to ensure a future for the world's coral reefs.

5

The Future of the Coral Reef

AS CORAL REEFS come under increasing pressure from human activities and warming temperatures, their survival is indeed in question. Toxic pollution, chemical runoff, sedimentation, coastal development, ship groundings, blast fishing, poison fishing, and other anthropogenic problems have all taken their toll on coral reefs around the globe. But in the last ten years, many people and organizations have realized the serious dangers that threaten coral reefs and are taking remedial action.

The United Nations declared 1997 as the Year of the Reef. Throughout the year, many conferences, events, and activities were organized to focus world awareness on the plight of coral reefs and to galvanize governments into protecting these valuable resources. Since that time, the momentum to protect coral reefs has continued. Coral reefs are still subject to natural and anthropogenic stresses. They are still bleaching, deteriorating, and dying. But around the world, action is being taken to try and reverse these trends and to save coral reefs and the marine species that inhabit them from irreversible damage.

The coral reef monitoring program

For many years, people have known that coral reefs were in trouble, but determining the specific problem, or how great the trouble, was difficult to tell. Because coral reefs are located under water, monitoring their health is difficult at best. Not only is proficient scuba diving essential, but scientific knowledge is also required for accurate assessment. In the early 1990s scientists began discussing the establishment of a global network that could monitor coral reefs everywhere.

In 1994, at a United Nations conference, the International Coral Reef Initiative (ICRI) was launched. Its purpose was to draw nations together with a renewed emphasis on monitoring coral reefs in their area of the globe. Since then, the ICRI secretariat and other organizations, such as the Australian Institute of Marine Science, have banded together and formed the Global Coral Reef Monitoring Network (GCRMN). In 1997, after establishing a subnetwork referred to as Reef Check, the network published its first annual State of the Reefs report.

Reef Check is the largest international coral reef monitoring program. According to the GCRMN,

> In 1997, Reef Check teams completed the first global survey of coral reefs. Over 750 volunteer sport divers were trained and led by 100 volunteer scientists in surveys of more than 300 reefs in 31 countries. The results of Reef Check 97 provided the first solid evidence that coral reefs have been damaged on a global scale.[30]

In particular, overfishing was shown to be far worse than expected, particularly in remote reef locations.

In 1998 forty countries participated in Reef Check. The results confirmed those of 1997: most reefs were severely overfished and many high-value organisms were missing. As the 1998 summary reports,

> Of the Worldwide Indicators, lobster were formerly abundant, and were missing from 85% of reefs surveyed. . . . There were no grouper [a large fish prized as food in Asian cultures] at 63% of reefs. Other Worldwide Indicators showed similar trends. . . . The reef corals themselves showed a decrease in living percentage of cover of more than 10%, and an increase in dead cover. A major cause of this change was an unprecedented "bleaching" event which killed many reef corals and other organisms, with the worst impacts in the Indo-Pacific.[31]

Climate Changes: Corals Tell the Tale

For millions of years coral reefs have been a fixture of the world's oceans. Reefs are also very sensitive to climate changes. Scientists know that, over millions of years, the earth's climate has changed and that coral have been affected by these changes. But now, coral itself may be able to tell the story.

Many corals lay down their skeletons in layers, or bands, as they grow, somewhat the way a tree develops a ring for each year of growth. These bands provide an accurate calendar record of climatic changes down to the exact year, season, and sometimes even the month or week the coral growth was deposited.

According to the National Oceanic and Atmospheric Administration (NOAA) in its February 18, 1997, report "How Has the Earth's Climate Changed,"

> Chemicals in the calcium carbonate skeleton can reveal invaluable information on ocean temperatures, rainfall/salinity, currents and upwelling. Examination of the skeletons can even reveal chemical contamination of their environment, the timing of their seasonal reproduction and past episodes of severe stress such as coral bleaching.

In obtaining an accurate record of climate changes, NOAA and other agencies are able to ascertain patterns and disturbances. For example, scientists can tell how coastal construction has changed the water flow around a reef or how silt or pollution has changed the reproduction cycles of polyps. The coral cores also provide information on whether the global warming currently being experienced is part of a naturally recurring cycle or is entirely driven by anthropogenic causes.

In addition to producing valuable scientific information, Reef Check has raised the awareness of divers, scientists, governments, and the general public about the value of coral reefs and the threats to their health. The results of both surveys have received worldwide television, newspaper, and magazine coverage in more than a dozen languages. Because Reef Check relies so heavily on local community involvement, it has been selected to be the community-based survey program for the United Nations Global Coral Reef Monitoring Network. Reef Check seeks to involve everyone and will be repeated each year. As Reef Check's 1999 information sheet begins, "If you are a scuba diver, marine scientist, community organizer or simply someone with a love of coral reefs, we need your help."

Although the survey results have confirmed the worst fears of most scientists, Reef Check and its parent network are optimistic about the long-term results. According to the GCRMN,

> Reef Check helps local community members learn how to monitor their coral reefs, providing the information needed so that they can be managed in a sustainable manner. Participation in Reef Check is one of the best methods of changing people's behavior and slowing the damage so that reefs can recover. Reef Check is one solution to the coral reef crisis.[32]

Many surfers and divers recognize the importance of maintaining healthy coral reefs.

Coral reefs and the diving and surfing communities

Divers and surfers were two of the first nonscientific community groups to recognize the importance of maintaining healthy coral reefs. For scuba divers, coral reefs are the preferred diving spots, and for surfers, the underwater presence of a coral reef creates the best surfing waves. To continue enjoying their recreational pastimes, divers and surfers realized that coral reefs need to remain healthy.

The Coral Reef Alliance (CORAL) has been instrumental in organizing divers and broadcasting the message of imperiled coral reefs around the globe. CORAL also actively promotes conservation by organizing reef

This mustard hill coral shows the dramatic effects of coral bleaching.

cleanups and awarding educational grants to marine sanctuaries and local communities. In 1998 CORAL awarded the Bonaire Marine Park a ten-thousand-dollar grant to be used for an outreach educational program. Bonaire is a small island in the Caribbean; the money will be used to teach island schoolchildren how to snorkel, scuba dive, and care for their magnificent reef. As Kalli de Meyer, Bonaire's park manager, explains, "The Bonaire Marine Park believes that the more the youth of Bonaire can see and learn about the beauty of coral reefs and the entire marine environment, the better stewards of the ocean they will be as they get older."[33] CORAL sponsors other similar programs around the globe.

Surfers, though admittedly a smaller population, are also becoming organized in the fight to save coral reefs. According to Pierce Flynn, the executive director of the Surfrider Foundation, "The best waves are often found breaking over thriving reefs. So we need to preserve reefs in order to preserve the sport."[34]

Surfers have taken a conservative approach to coastal development in areas bordering coral reefs. For example, the coral reefs that line Indonesia's coastline provide some of the best surfing in the world. Surfers come from all over the globe just to surf in Indonesia, often for weeks at a time. All of these surfers need food and lodging, but the surfing community has been nearly unanimous in not wanting these areas developed with high-rise hotels, condominiums, and restaurants.

At Grajagan Bay, Indonesia, two rustic surf camps have been built along the shoreline next to a nature reserve. Both have been designed and are operated to have no significant environmental impact. In this instance, surfers from around the world have provided the economic incentive to preserve and maintain the coral reefs. Additionally, Quiksilver, an Australian-based company that is the world's largest manufacturer of surf clothing, also sponsors an annual surfing contest at Grajagan Bay and actively promotes conservation of the area. Quiksilver has used its corporate strength to convince the Indonesian government to conserve the area and also to assure that people of the area are able to make a living without destroying the coral reefs.

Surfing has provided countries that are home to coral reefs with an economic incentive to protect their waters.

Coral and the marine aquarium trade

The marine aquarium trade delivers a double whammy to coral reefs. In gathering specimens for sale, the industry usually removes the largest and/or best species from a reef, and in gathering them, cyanide and other poisons are usually used. To create appropriate habitat for tropical marine aquariums, live coral is also collected by breaking pieces off the reef, often using crowbars.

Although the marine aquarium trade certainly contributes to the degradation of coral reefs, in the last few years it has been the marine aquarium trade community that has been most active in the search for solutions. Realistically, it is the marine aquarium trade that has one of the largest interests in maintaining healthy reefs: If the coral reefs die, there will be no marine aquarium trade.

The Marine Aquarium Fish Council (MAFC) is in the process of developing a program to certify marine ornamental specimens. Only fish and invertebrates that have been collected in a sustainable manner (without using poison or other reef-destructive methods) will receive the MAFC certification. The MAFC is hoping to educate consumers not to buy fish that lack the certification. Through this process, the MAFC is also hoping to make it economically impractical to collect coral reef specimens in a destructive manner because no one will buy them.

The Trade Advisory Group (TAG) has also been implementing training programs through member organizations such as the International Marinelife Alliance and the Federation of Fish Collectors of the Philippines. In addition, TAG is working with other organizations to train local fishermen to use methods other than cyanide when capturing specimens.

Coral farming

Because coral is a living organism, theoretically it should be possible to harvest it like other renewable resources, as long as the demand for live coral doesn't outstrip the coral's natural ability to regenerate and grow. Unfortunately, the demand for live coral in marine aquariums, combined with

Public Aquariums: Educating and Conserving

It's not necessary to have scuba equipment or a degree in marine biology to learn about coral reefs or experience their wonders. Public aquariums throughout the world are developing informative and fascinating coral reef programs. Interactive coral exhibits can be found at aquariums in Florida, California, Maryland, Illinois, and Hawaii, among other places.

In Chicago, the Shedd Aquarium has its own ninety-thousand-gallon tropical reef, complete with sea turtles, sharks, moray eels, and a diver who hand-feeds them. Additionally, there is an interactive exhibit showing the life cycle of a coral reef—from flourishing, to stripped of life.

In Honolulu, the Waikiki Aquarium has maintained a coral exhibit since 1978, with seventy-four different species on public display. It also offers a coral propagation exhibit that is open to the public.

Aquariums provide many people with the opportunity to witness the supreme beauty of coral reefs.

all of the other stresses that coral is subject to, have made the collection of live coral an unsustainable practice, just as cyanide fishing has become an unsustainable practice.

Some scientists and aquariums have attempted to raise coral in captivity. By propagating "domestic" coral, they are hoping to take some of the pressure off wild coral. The Waikiki Aquarium at the University of Hawaii in Honolulu and Sea World of Ohio are two facilities that are successfully raising coral in captivity, albeit on a very small scale.

With a recent twenty-thousand-dollar grant from the Conservation Endowment Fund of the American Zoo and Aquarium Association (AZA), the Waikiki Aquarium was able to double the size of its coral farm and make more cultured corals available to public aquariums and researchers. The new seventeen-hundred-gallon facility is open to the public, and graphics and informational signs educate visitors about coral biology and the techniques used in raising coral in captivity.

According to a fact sheet put out by NOAA,

> The Waikiki Aquarium maintains 74 species of stony corals from Fiji, Solomon Islands, Palau, Guam, and Hawaii (some dating back to 1980). Because corals can grow seven inches or more per year, they require periodic "pruning." The fragments, or "cuttings," generated by this pruning activity are made available to other public aquariums and researchers.[35]

Sea World of Ohio has been so successful in its coral farming program that it is able to supply all of the other Sea World parks in California, Florida, and Texas with live coral fragments for their marine displays. By the mid-1990s, Sea World of Ohio had doubled the biomass of its original coral collection, simply due to the rapid growth of two-inch live cuttings, or prunings, into grapefruit-sized individual colonies.

Through the success of these coral farming operations, scientists are hoping to increase the supply of farmed coral commercially available to the home aquarist, thereby further lowering the demand for wild coral. By some estimates, 20 percent of today's live coral trade is already supplied through coral farming.

Coral reefs and the fishing industry

One of the greatest anthropogenic dangers to coral reefs remains the reef-destructive fishing practices used by many Pacific and Asian fishermen. Overfishing, through regular, cyanide, and blast methods, remains a persistent problem.

Community-managed coral reef education and conservation programs appear to have the best chance of succeeding. If the local community has a stake in the health of its coral reefs, the reefs have a better chance of remaining healthy. In Indonesia, at the request of the Indonesian Ministry of Forestry, the Nature Conservancy (TNC) has developed a program to assist the Komodo National Park in protecting the park's marine ecosystem. Throughout the park, the coral reefs ringing the islands have been heavily damaged by blast fishing. In 1996 TNC established a field office in the park and began working with local villagers to develop a program to address destructive fishing practices.

Komodo National Park has been especially successful in reducing blast fishing within its borders.

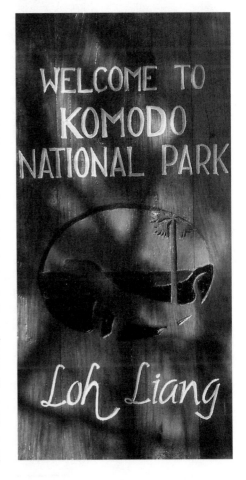

According to NOAA, "To deal with blast fishing, TNC helped form an enforcement team consisting of [park] managers, police, and fishermen. Since the team's inception, in 1995, blast fishing has declined by more than 80%."[36]

For the more than 150,000 people living along the Tanga coast of northeast Tanzania in Africa, the surrounding coral reefs provide the main source of income and protein. Healthy reefs are vital to the health of the area. Unfortunately, much of the reef structure along this 180-kilometer section of coast has been destroyed by blast fishing and other destructive fishing practices. Because the Tanzanian government has found it difficult to enforce regulations, it has developed a community-based education and enforcement program. The program is considered innovative because all villagers have a stake in the success of the program.

After two community-wide meetings, three villages created committees to deal with the problem of reef-destructive fishing, the increasing demand for fish, and the increasing numbers of fishermen in the area. Through community action, certain types of fishing are controlled, and sections of the reef are closed to any kind of harvesting. Village elders inspect fishing gear and the catch of local fishermen as well as patrol the local markets inspecting what fish are being offered for sale. There are also patrols against fishermen using dynamite.

So far, the efforts of the three villages have been successful: More fish are actually being caught and the reef is recovering. More villages have come forward, asking to be part of the program. According to NOAA, "The demand for community-based management programs has exceeded the government's ability to respond, and demonstrated to the government the advantages of involving coastal communities in coral reef management."[37]

National and international efforts

In the United States, the continued degradation of coral reefs poses a serious threat to U.S. fisheries, with possible annual losses in the hundreds of millions of dollars. According to NOAA,

> There are approximately 500 federally managed species of fish and marine invertebrates that depend on coral reefs during part of their life cycle. Red snapper, . . . grouper, . . . and spiny lobster are just some of the important commercial species that may not survive, and certainly will not thrive, without healthy coral reefs.[38]

Additionally, coral reefs also provide coastal communities with millions of dollars in recreational and tourism activities every year. Consequently, the United States has taken some action on its own to protect these valuable resources.

With the passage of the Marine Protection, Research, and Sanctuaries Act and the Coastal Zone Management Act in 1972, the United States took its first steps toward protecting coral reefs. Since that time, twelve marine sanctuaries have been established in U.S. waters, including coral reefs in the Florida Keys, Hawaii, the Gulf of Mexico, and American Samoa.

In 1997 the U.S. House of Representatives introduced H.R. 2233, the Coral Reef Conservation Act. It was designed to help preserve coral reefs by establishing conservation programs and providing financial support for those programs. Although H.R. 2233 passed the House, it failed to pass in the Senate. As of June 1999, the bill was still being reworked in hopes of reintroducing it in a future session.

On the international front, the World Conservation Union has completed proposals and funding to develop marine sanctuaries in Tanzania, Western Samoa, and Vietnam. The United Nations is also involved in the preservation of coral reefs around the globe, through launching and/or supporting such programs as the International Coral Reef Initiative, Reef Check, and others.

A marine patrolman helps to protect the reefs at Lode Key National Marine Sanctuary.

The Convention on International Trade in Endangered Species of Wild Fauna and Flora (CITES) has also become involved in the plight of coral reefs. CITES attempts to assign all plants and animals to an appendix category. A plant or animal that is listed on Appendix I cannot be traded at all. A plant or animal that is listed on Appendix II may be traded internationally, but the exporting country must show documentation. Black, blue, stony, fire, and organ-pipe corals are all listed on Appendix II and must have a permit to be exported or sold.

Although coral trade is restricted, it is difficult to monitor the size of the actual trade and its impact on existing coral reef ecosystems. According to a recent CITES report, Indonesia currently supplies 95 percent of the world's coral trade, and the United States imports 85 percent of all dead coral and 98 percent of all live coral that is internationally traded. As more information is obtained through Reef Check, the CITES regulations and classifications may tighten.

Coral reefs in the next millennium

Global warming still remains the greatest natural and anthropogenic stress that corals are currently experiencing—and the most daunting to confront. Even if global warming is caused entirely by human activities and nations around the globe agree immediately on a plan of action, implementing a conservation program and measuring results is a decades-long project.

But every conservation effort helps coral reefs. According to CORAL's Summer 1998 newsletter,

> While CORAL's conservation programs cannot prevent global warming or other climatic threats to coral reefs, they can help minimize the impact of bleaching on the world's reefs. Local conservation efforts—such as Marine Protected Areas that CORAL helps establish—reduce the stress to coral reef ecosystems by reducing: over-fishing, excess nutrients, sediment and other environmental threats to coral reefs. When these threats are reduced, the corals are better able to recover and recolonize.[39]

When all of these conservation efforts are combined, coral reefs stand a chance to recover. If the only stress they encounter is rising water temperatures—if they don't have to fight off all other anthropogenic stresses—they may be able to withstand global warming. With continuing conservation and public awareness, perhaps these small animals that have created magnificent ecosystems for millions of years will be around for millennia to come.

Notes

Chapter 1: A Living Animal

1. Thomas M. Niesen, *The Marine Biology Coloring Book.* New York: Barnes & Noble Books, 1982, p. 59.
2. Niesen, *The Marine Biology Coloring Book,* p. 59.
3. Rebecca L. Johnson, *The Great Barrier Reef: A Living Laboratory.* Minneapolis: Lerner, 1991, p. 21.
4. Niesen, *The Marine Biology Coloring Book,* p. 59.
5. Niesen, *The Marine Biology Coloring Book,* p. 59.
6. James L. Sumich, *An Introduction to the Biology of Marine Life,* 6th ed. Dubuque, IA: Wm. C. Brown, 1996, p. 255.
7. Sumich, *An Introduction to the Biology of Marine Life,* p. 258.
8. Sumich, *An Introduction to the Biology of Marine Life,* p. 259.
9. J. E. N. Veron, *Corals of Australia and the Indo Pacific.* Honolulu: University of Hawaii Press, 1993, p. 46.

Chapter 2: An Underwater City

10. Douglas H. Chadwick, "Coral in Peril," *National Geographic,* January 1999, p. 32.
11. Niesen, *The Marine Biology Coloring Book,* p. 9.
12. Sumich, *An Introduction to the Biology of Marine Life,* p. 261.
13. Sumich, *An Introduction to the Biology of Marine Life,* p. 261.

Chapter 3: Environmental Threats to Coral Reefs

14. National Oceanic and Atmospheric Administration, "Corals Can't Stand the Heat: Global Climate Change and Coral Reefs," April 28, 1997. www.noaa.gov/public-affairs/iyorwk17.html.
15. National Oceanic and Atmospheric Administration,

"Corals Can't Stand the Heat: Global Climate Change and Coral Reefs."

16. Quoted in National Oceanic and Atmospheric Administration, "Florida Keys Bleaching Event May Be Linked to El Niño," September 1, 1997. www.noaa.gov/public-affairs/iyorwk 35.html.

17. Quoted in National Oceanic and Atmospheric Administration, "Florida Keys Bleaching Event May Be Linked to El Niño."

18. National Oceanic and Atmospheric Administration, "Coral Reefs and Heavy Metals: Land-Based Threats, Part I," July 28, 1997. www.noaa.gov/public-affairs/iyorwk30.html.

19. National Oceanic and Atmospheric Administration, "Coral Reefs and Heavy Metals: Land-Based Threats, Part I."

20. Quoted in National Oceanic and Atmospheric Administration, "Coral Reefs Under Stress from Double Whammy: Sediments and Nutrients," September 29, 1997. www.noaa.gov/public-affairs/iyorwk39.html.

21. Coral Reef Alliance (CORAL), "African Dust May Cause Deterioration of Coral Reefs," Reef Updates Around the World, Winter 1999 newsletter. www.coral.org.

22. Coral Reef Alliance, "African Dust May Cause Deterioration of Coral Reefs."

23. Quoted in National Oceanic and Atmospheric Administration, "Coral Reefs and the Coastal Zone," July 14, 1997. www.noaa.gov/public-affairs/iyorwk28.html.

Chapter 4: Fishing and Recreational Threats to Coral Reefs

24. Stephen C. Jameson, John W. McManus, and Mark D. Spalding, "Executive Summary, State of the Reefs Report," in "State of the Reefs, Regional and Global Perspectives," May 1995. International Coral Reef Initiative Executive Secretariat. www.ogp.noaa.gov/misc/coral/sor/sor_exsum.html.

25. Coral Reef Alliance, "Fishing with Cyanide = Coral Reef Genocide," 1999. www.coral.org.

26. National Oceanic and Atmospheric Administration, "Poisons Used to Harvest Fish Are Killing World's Reefs," March 10, 1997. www.noaa.gov/public-affairs/iyorwk10.html.

27. National Oceanic and Atmospheric Administration, "When Coral Reefs Get in the Way," April 21, 1997. www. noaa.gov/public-affairs/iyorwk16.html.

28. National Oceanic and Atmospheric Administration, "Scuba Divers and Coral Reefs: A Natural Alliance," September 15, 1997. www.noaa.gov/public-affairs/iyorwk 37.html.

29. Coral Reef Alliance, "About the Coral Reef Alliance," 1999. www.coral.org.

Chapter 5: The Future of the Coral Reef

30. Global Coral Reef Monitoring Network, "1999 Reef Check," 1999. www.reefcheck.org.

31. Global Coral Reef Monitoring Network, "Reef Check 1998 Summary Results," 1999. www.reefcheck.org.

32. Global Coral Reef Monitoring Network, "1999 Reef Check."

33. Coral Reef Alliance, "A CORAL Present for Bonaire's Future," CORAL News, Summer 1998. www.coral.org.

34. Quoted in National Oceanic and Atmospheric Administration, "The Wave of the Future: Surfing and Coral Reef Conservation," May 26, 1997. www.noaa.gov/public-affairs/iyorwk21.html.

35. National Oceanic and Atmospheric Administration, "Coral Farming Takes Pressure off Coral Reefs," July 21, 1997. www.noaa.gov/public-affairs/iyorwk29.html.

36. National Oceanic and Atmospheric Administration, "Biodiverse Coral Reefs of Komodo National Park Threatened by Reef-Destructive Fishing," August 4, 1997. www.noaa.gov/public-affairs/iyorwk31.html.

37. National Oceanic and Atmospheric Administration, "Community-Based Management Saving the Coral Reefs of Tanzania," June 23, 1997. www.noaa.gov/public-affairs/iyorwk25.html.

38. National Oceanic and Atmospheric Administration, "Plight of Coral Reefs a Threat to Lucrative U.S. Fisheries," August 18, 1997. www.noaa.gov/public-affairs/iyorwk33.html.

39. Coral Reef Alliance, "CORAL News."

Glossary

Atoll reef: The oldest type of coral reef. Atolls are ring-shaped reefs that surround a lagoon.

Back reef: The section of reef that faces the lagoon.

Barrier reef: The second oldest type of coral reef. Barrier reefs are separated from the shoreline of an island or landmass by a lagoon.

Bleaching: The process by which a coral polyp expels its zooxanthellae.

Budding: The process of asexual reproduction whereby a coral polyp "buds" off a "daughter" and replicates itself.

Calcium: The primary mineral component of a corallite.

Coelenterate: The taxonomic classification of the phylum of animals that includes coral, anemones, and jellyfish.

Coenosarc: The thin connective tissue that grows between polyps on a reef.

Coral: A tiny invertebrate in the coelenterate phylum.

Corallite: The hard skeletal cup, or exoskeleton, in which an adult coral polyp lives. Each polyp makes its own corallite by secreting a substance composed primarily of calcium.

Egg bundle: A small ball of reproductive cells produced by a coral polyp.

Exoskeleton: A skeleton, or skeletal structure, that is on the outside of an animal's body.

Fringing reef: The youngest type of coral reef. Fringing reefs extend seaward, directly from the shoreline of an island or landmass.

Hard coral: Any species of coral that forms a stony exoskeleton.

Invertebrate: An animal without a backbone; marine invertebrates include crabs, octopuses, and coral.

Lagoon: A shallow body of water that exists between a shoreline and a barrier reef.

Lower reef slope: The portion of the reef front that is in deeper water, receives less sunlight, and slopes downward until it reaches the seabed.

Photosynthesis: The process of a plant obtaining nourishment and creating energy from sunlight.

Plankton: Small animals and plants, often microscopic single-celled organisms, that live in the ocean and form the foundation of the marine food chain.

Planula: A coral in the infant stage.

Polyp: An adult coral.

Reef crest: The highest point of the reef. The reef crest divides the back reef from the reef front.

Reef flat: A long flat area over the lagoon that leads to the reef crest.

Reef front: The part of the reef that faces outward toward the open ocean.

Soft coral: Any species of coral that does not have a hard exoskeleton.

Spawning: The process of releasing eggs and sperm into the water so that sexual reproduction can occur. Most marine creatures reproduce sexually through spawning.

Symbiosis: A relationship between two or more organisms that benefits all members.

Upper reef slope: The portion of the reef front that is closest to the reef crest and receives the most sunlight.

Zooxanthellae: Microscopic algae that can be found in the ocean and can also be found within coral polyps.

Organizations to Contact

American Zoo and Aquarium Association
8403 Colesville Rd.
Silver Spring, MD 20910
(301) 562-0777
fax: (301) 562-0888
e-mail: zoognus@aol.com
website: www.aza.org

An organization that represents 180 accredited zoos and aquariums in North America and can provide information about any of them.

Coral Reef Alliance (CORAL)
64 Shattuck Sq., Suite 220
Berkeley, CA 94794
(510) 848-0110
fax: (510) 848-3720
website: www.coral.org

An international nonprofit membership association of scuba divers and other concerned individuals who work to address the worldwide problem of coral reef destruction.

Great Barrier Reef Marine Park Authority
PO Box 1379
Townsville QLD 4810, Australia
61-7-4750-0700
fax: 61-7-4772-6093
website: www.gbrmpa.gov.au

The agency of the Australian government that manages the Great Barrier Reef. Its website is full of information about the Great Barrier Reef, specifically, and coral reefs in general.

National Oceanic and Atmospheric Administration
U.S. Department of Commerce
14th St. and Constitution Ave., NW, Room 6013
Washington, DC 20230
(202) 482-6090
fax: (202) 482-3154
website: www.noaa.gov/

The agency within the U.S. government that is responsible for all national and international management of coral reefs. Its website offers hundreds of information and fact sheets and links to other sites regarding coral reefs. One of its additional websites, www.noaa.gov/coral_links.html, offers nine pages of additional coral link sites.

World Conservation Union
Marine and Coastal Programme
Rue Mauvernay 28
CH 1196 Gland
Switzerland
61-6-2511-402
fax: 61-6-2475-761

An international organization devoted to the establishment of protected marine sanctuaries and conservation programs around the world. The Marine and Coastal Programme is just one of many areas that the union pursues.

World Resources Institute
10 G St., NE, Suite 800
Washington, DC 20002
(202) 729-7600
fax: (202) 729-7610
website: www.wri.org/wri

An organization that works to integrate "natural resource use and conservation, economic development, and social equity through research, capacity building and institutional change." It offers many publications on coral reefs.

Suggestions for Further Reading

Rachel Carson, *The Sea Around Us.* New York: Penguin Books, 1950. A very readable classic that portrays the wonders of the ocean ecosystem.

J. Y. Cousteau with James Dugan, *The Living Sea.* New York: Harper & Row, 1963. One of the world's most famous oceanographers recounts his diving experiences, most of them on coral reefs.

Carl Roessler, *Coral Kingdoms.* New York: Harry N. Abrams, 1986. A large-format, interesting book illustrated with full-color photographs.

Charles R. C. Sheppard, *A Natural History of the Coral Reef.* Dorset, England: Blandford, 1983. A well-illustrated, informative book that deals with all aspects of a coral reef.

Robert Silverberg, *The World of Coral.* New York: Duell, Sloan, and Pearce, 1965. A readable book that covers corals themselves as well as life on a coral reef.

Works Consulted

Books

Rebecca L. Johnson, *The Great Barrier Reef: A Living Laboratory.* Minneapolis: Lerner, 1991. A comprehensive overview of the world's largest barrier reef.

Thomas M. Niesen, *The Marine Biology Coloring Book.* New York: Barnes & Noble Books, 1982. Not really a coloring book at all, but rather a complete marine biology textbook illustrated in black and white.

James L. Sumich, *An Introduction to the Biology of Marine Life.* 6th ed. Dubuque, IA: Wm. C. Brown, 1996. A detailed college-level textbook covering all aspects of marine biology.

J. E. N. Veron, *Corals of Australia and the Indo Pacific.* Honolulu: University of Hawaii Press, 1993. A detailed graduate-level textbook covering different coral species of the stated region.

Periodicals

Douglas H. Chadwick, "Coral in Peril," *National Geographic,* January 1999.

David Doubliet, "Coral Eden," *National Geographic,* January 1999.

New York Times, "As Oceans Warm, Problems from Viruses and Bacteria Mount," January 24, 1999.

Websites

Coral Reef Alliance (CORAL), "About the Coral Reef Alliance," 1999. www.coral.org.

Coral Reef Alliance, "African Dust May Cause Deterioration of Coral Reefs," Reef Updates Around the World, Winter 1999 newsletter. www.coral.org.

Coral Reef Alliance, "Coral News: A CORAL Present for Bonaire's Future," Summer 1998. www.coral.org.

Coral Reef Alliance, "Fishing with Cyanide = Coral Reef Genocide," 1999. www.coral.org.

Global Coral Reef Monitoring Network, "An Overview of the Global Coral Reef Monitoring Network," 1999. www.coral.noaa.gov/gcrmn.html.

Great Barrier Reef Marine Park Authority, "COTSWATCH—International," 1999. www.gbrmpa.gov.au/cots.

Stephen C. Jameson, John W. McManus, and Mark D. Spalding, "State of the Reefs, Regional and Global Perspectives," May 1995. International Coral Reef Initiative Executive Secretariat. www.ogp.noaa.gov/misc/coral/sor/sor_exsum.html.

National Oceanic and Atmospheric Administration, "Biodiverse Coral Reefs of Komodo National Park Threatened by Reef-Destructive Fishing," August 4, 1997. www.noaa.gov/public-affairs/iyorwk31.html.

National Oceanic and Atmospheric Administration, "CITES and the International Coral Trade," June 2, 1997. www.noaa.gov/public-affairs/iyorwk22.html.

National Oceanic and Atmospheric Administration, "Community-Based Management Saving the Coral Reefs of Tanzania," June 23, 1997. http://www.noaa.gov/public-affairs/iyorwk25.html.

National Oceanic and Atmospheric Administration, "Coral Farming Takes Pressure off Coral Reefs," July 21, 1997. www.noaa.gov/public-affairs/iyorwk29.html.

National Oceanic and Atmospheric Administration, "A Coral Reef in Your Medicine Cabinet: The Biomedical Applications of Marine Organisms," March 31, 1997. www.noaa.gov/public-affairs/iyorwk13.html.

National Oceanic and Atmospheric Administration, "Coral Reefs and Heavy Metals: Land-Based Threats, Part I," July 28, 1997. www.noaa.gov/public-affairs/iyorwk30. html.

National Oceanic and Atmospheric Administration, "Coral Reefs and the Coastal Zone," July 14, 1997. www.noaa.gov/public-affairs/iyorwk28.html.

National Oceanic and Atmospheric Administration, "Coral Reefs Under Stress from Double Whammy: Sediments and Nutrients," September 29, 1997. www.noaa.gov/public-affairs/ iyorwk39.html.

National Oceanic and Atmospheric Administration, "Corals Can't Stand the Heat: Global Climate Change and Coral Reefs," April 28, 1997. www.noaa.gov/public-affairs/ iyorwk17.html.

National Oceanic and Atmospheric Administration, "Florida Keys Coral Bleaching Event May Be Linked to El Niño," September 1, 1997. www.noaa.gov/public-afairs/iyorwk35. html.

National Oceanic and Atmospheric Administration, "How Has the Earth's Climate Changed? Coral Cores May Tell the Tale," February 18, 1997. www.noaa.gov/public-affairs/iyorwk7.html.

National Oceanic and Atmospheric Administration, "Plight of Coral Reefs a Threat to Lucrative U.S. Fisheries," August 18, 1997. www.noaa.gov/public-affairs/iyorwk33.html.

National Oceanic and Atmospheric Administration, "Poisons Used to Harvest Fish Are Killing World's Reefs," March 10, 1997. www.noaa.gov/public-affairs/iyorwk10.html.

National Oceanic and Atmospheric Administration, "Scuba Divers and Coral Reefs: A Natural Alliance," September 15, 1997. www.noaa.gov/public-affairs/iyorwk37.html.

National Oceanic and Atmospheric Administration, "The Wave of the Future: Surfing and Coral Reef Conservation," May 26, 1997. www.noaa.gov/public-affairs/iyorwk21.html.

National Oceanic and Atmospheric Administration, "When Coral Reefs Get in the Way," April 21, 1997. www.noaa.gov/public-affairs/iyorwk16.html.

Reef Check, "Reef Check 1998 Summary Results." www.reefcheck.org.

Reef Check, "1999 Reef Check." www.reefcheck.org.

Index

Picture Credits

About the Author

Lesley A. DuTemple is the author of many natural history books for children and young adults, covering such topics as tigers, whales, polar bears, moose, and others. DuTemple lives on the edge of a canyon in Salt Lake City, Utah. She, her husband, and two young children share their property with a family of raccoons, a resident porcupine, several flocks of quail and songbirds, two peregrine falcons, and roaming herds of deer.